AF761934

MECHANICAL DESIGN OF PIEZOELECTRIC ENERGY HARVESTERS

MECHANICAL DESIGN OF PIEZOELECTRIC ENERGY HARVESTERS

Generating Electricity from Human Walking

QINGSONG XU

Department of Electromechanical Engineering, Faculty of Science and Technology, University of Macau, Avenida da Universidade, Macau, China

LAP MOU TAM

Department of Electromechanical Engineering, Faculty of Science and Technology, University of Macau, Avenida da Universidade, Macau, China

ELSEVIER

ACADEMIC PRESS

An imprint of Elsevier

Academic Press is an imprint of Elsevier
125 London Wall, London EC2Y 5AS, United Kingdom
525 B Street, Suite 1650, San Diego, CA 92101, United States
50 Hampshire Street, 5th Floor, Cambridge, MA 02139, United States
The Boulevard, Langford Lane, Kidlington, Oxford OX5 1GB, United Kingdom

Copyright © 2022 Elsevier Inc. All rights reserved.

No part of this publication may be reproduced or transmitted in any form or by any means, electronic or mechanical, including photocopying, recording, or any information storage and retrieval system, without permission in writing from the publisher. Details on how to seek permission, further information about the Publisher's permissions policies and our arrangements with organizations such as the Copyright Clearance Center and the Copyright Licensing Agency, can be found at our website: www.elsevier.com/permissions.

This book and the individual contributions contained in it are protected under copyright by the Publisher (other than as may be noted herein).

Notices
Knowledge and best practice in this field are constantly changing. As new research and experience broaden our understanding, changes in research methods, professional practices, or medical treatment may become necessary.

Practitioners and researchers must always rely on their own experience and knowledge in evaluating and using any information, methods, compounds, or experiments described herein. In using such information or methods they should be mindful of their own safety and the safety of others, including parties for whom they have a professional responsibility.

To the fullest extent of the law, neither the Publisher nor the authors, contributors, or editors, assume any liability for any injury and/or damage to persons or property as a matter of products liability, negligence or otherwise, or from any use or operation of any methods, products, instructions, or ideas contained in the material herein.

Library of Congress Cataloging-in-Publication Data
A catalog record for this book is available from the Library of Congress

British Library Cataloguing-in-Publication Data
A catalogue record for this book is available from the British Library

ISBN: 978-0-12-823364-1

For information on all Academic Press publications visit our website at https://www.elsevier.com/books-and-journals

Publisher: Mara Conner
Acquisitions Editor: Carrie Bolger
Editorial Project Manager: Joshua Mearns
Production Project Manager: Prem Kumar Kaliamoorthi
Cover Designer: Miles Hitchen

Typeset by TNQ Technologies

Working together to grow libraries in developing countries
www.elsevier.com • www.bookaid.org

To our families.
For the celebration of the 40th anniversary of the University of Macau.

Contents

Preface

With the rapid advances in microelectrical-mechanical systems (MEMS), developments in MEMS sensors, actuators, and controllers have greatly improved. Traditional energy supply devices such as batteries have weaknesses in terms of low energy density, a short working life, and a heavy maintenance workload. Hence, they cannot satisfy the demand of MEMS power supplies. Alternatively, as a kind of self-sustaining energy supply device, a microenergy harvester offers a new energy supply. In particular, piezoelectric materials can convert mechanical energy into electric energy based on the direct piezoelectric effect. Piezoelectric energy harvesters are presented to harvest energy from the ambient environment, such as human motion, vibration, and flow. Such devices are beneficial to the deployment of Internet of Things connections for the sustainable development of smart cities worldwide.

Capturing energy from human walking to replace or supplement batteries has drawn the great attention of researchers owing to the presence of the significant amount of harvestable energy in walking motion compared with other types of human movement. Generally, piezoelectric energy harvesters have the merits of a small footprint, simple structure, and high energy density. However, conventional piezoelectric energy harvesters have low energy conversion efficiency and can scavenge only a very low level of power. It is challenging to devise an efficient piezoelectric energy harvester with a high operating bandwidth, large output power, and safe structure for specific applications.

This book is focused on enabling technologies in the design and development of piezoelectric energy harvesters for generating energy from human walking. It covers mechanical designs with a compliant mechanism, piezoelectric stack, and permanent magnet for high power—output generation tasks. The book proposes innovative mechanism designs of one-, two-, and three-dimensional piezoelectric energy harvesters along with experimental verifications. The book also presents designs for force-amplification compliant mechanisms and force direction-transmission mechanisms dedicated to scavenging energy from human walking footsteps. Comprehensive treatment of the subject matter is provided in a manner amenable to readers ranging from researchers to engineers, by supplying detailed experimental verifications of the developed devices.

This book is composed of 12 chapters. The book begins with an introduction to energy harvesting, and offers a brief survey of development in piezoelectric energy harvesters for applications generating energy from the ambient environment. Chapter 2 presents a comprehensive review of state-of-the-art developments in energy harvesting devices from human walking. Chapter 3 outlines advances in mechanical designs of piezoelectric type of energy harvesters, in which the current development progress and remaining challenges are discussed.

Regarding the one-dimensional energy harvester, Chapter 4 devises a piezoelectric energy harvester based on a compound two-stage force amplification frame. Chapter 5 proposes a dual-mode energy harvesting handrail to generate energy from both vibrations and pulling force excitation modes. Furthermore, Chapter 6 presents a novel piezoelectric energy harvester using an integrated multistage force amplification mechanism to scavenge mechanical energy from human footsteps during walking.

Generally, an energy harvester is more feasible if it can accept input excitation in more directions. Chapter 7 designs a novel piezoelectric energy harvester with a single piezoelectric stack to harvest energy from excitations with bidirectional motion. Chapter 8 devises a piezoelectric energy harvester to harvest energy from two-degrees-of-freedom excitation in the vertical direction using a single piezoelectric stack. Chapter 9 develops a two-dimensional piezoelectric energy harvester with permanent magnets and a multistage force amplification frame. In addition, Chapter 10 demonstrates a dual-axial underfloor piezoelectric energy harvester based on a piezoelectric stack.

Concerning a three-dimensional energy harvester design, Chapter 11 proposes the concept design of a novel piezoelectric energy harvester to harvest energy from three-degrees-of-freedom force excitation using a single piezoelectric stack and several force transmission mechanisms. To broaden the frequency bandwidth of the harvester, Chapter 12 reports a novel bistable piezoelectric energy harvester dedicated to vibration energy harvesting in the direction of gravity.

This book provides state-of-the-art emerging techniques to develop piezoelectric energy harvesters to generate electric energy from human walking. It covers the topics of compliant mechanisms, mechanism design, piezoelectric materials, analytical modeling, optimum designs, simulation studies, and experimental investigations. Detailed examples of their implementations are provided. Readers can expect to learn how to design and develop new piezoelectric energy harvesting devices to scavenge energy from multidimensional vibration and force excitations.

Qingsong Xu and Lap Mou Tam
Macau, China

Acknowledgements

The authors would like to acknowledge the Science and Technology Development Fund of Macao SAR (under Grants 143/2016/A and 0022/2019/AKP) and the Research Committee of the University of Macau (under Grant MYRG2019-00133-FST) for co-funding the projects.

CHAPTER 1
Introduction

This chapter provides an introduction to the design and application of piezoelectric energy harvesting devices. The motivation for employing an energy harvester in practice is discussed. The advantages of piezoelectric energy harvester are presented. In particular, the mechanical design of a piezoelectric energy harvester for scavenging energy from human walking is outlined.

1.1 Why energy harvesting is used in practice

With the development of the Internet of Things (IoT), various sensors have been widely deployed in our daily lives, providing big data via the IoT. For example, traffic sensors can be embedded into roads to monitor the status of traffic (Nellore & Hancke, 2016; Souri et al., 2015). Hard wiring has the shortcomings of high cost and complex cabling, but these issues can be overcome with the maturation and standardization of wireless networks. For instance, portable electronic devices such as smartphones and watches are common in our everyday lives. Widely deployed sensors will penetrate every industry to enable data collection. Such devices feature low power consumption and a compact footprint size.

The main power sources for sensors are rechargeable batteries. However, the periodic recharge process of batteries and low energy density are limiting factors that restrict the mobility of devices. More important, the pollution of waste batteries has a serious impact on the global environment. The power requirement for microelectrical-mechanical systems (MEMS) devices (e.g., wireless sensors) is reduced to microwatts or nanowatts. It creates the possibility of replacing current batteries with new power-supplying devices that can reduce dependence on external power sources. Hence, the development of alternative energy sources with low maintenance and a long lifetime has become the subject of extensive study (Nersesian, 2014).

Significant interest has been paid to energy harvesters that can convert various forms of energy in an ambient environment, such as wind, sunshine,

Mechanical Design of Piezoelectric Energy Harvesters
ISBN 978-0-12-823364-1
https://doi.org/10.1016/B978-0-12-823364-1.00007-9
© 2022 Elsevier Inc.
All rights reserved.

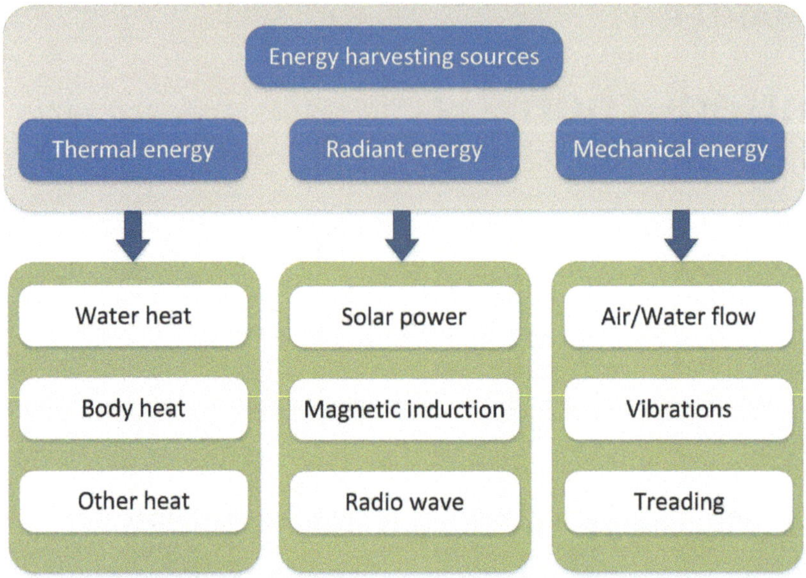

Figure 1.1 Classification of energy harvesting sources.

and vibration, into electric energy (He et al., 2021; Wang et al., 2021; Zou et al., 2021). A classification of energy harvesting sources is shown in Fig. 1.1. In particular, because of increasing demands for low-power supply, vibration-type mechanical energy harvesters have become popular because vibration sources are ubiquitous. Main kinds of vibration energy harvesting devices involve electromagnetic, electrostatic, and piezoelectric type transducers (Du et al., 2018; Klein & Zuo, 2017; Liao & Liang, 2018).

1.2 Piezoelectric energy harvester

Piezoelectric energy harvesters have become a popular research topic (Covaci & Gontean, 2020). The working principle of piezoelectric materials is based on the transformation of mechanical energy provided by the input force into electrical energy, which is called the direct piezoelectric effect. The piezoelectric material generates electricity when it experiences a vibration or press/release force. Because it can transform mechanical energy from the ambient environment into electrical energy, it serves as a kind of self-sustaining energy-supply device.

Around 200 piezoelectric materials have been applied in various energy harvesting applications (Mishra et al., 2018; Priya et al., 2017). The materials exhibit different properties in terms of physical size, piezoelectric coefficients, acoustic impedance, and so on (Zhengbao & Jean, 2016). These piezoelectric materials can be classified into four categories: (1) single crystals (e.g., Rochelle salt, lithium niobite, and quartz crystals); (2) ceramics (e.g., lead-zirconate-titanate [PZT], barium titanate ($BaTiO_3$), and potassium niobate [$KNbO_3$]); (3) polymers (e.g., polyvinylidene fluoride [PVDF], polylactic acid [PLA], and copolymers); and (4) polymer composites or nanocomposites (e.g., PVDF-zinc oxide [PVDF-ZnO], polyimide-PZT, and cellulose $BaTiO_3$).

Because of ultralow power consumption (in nano- or microwatts), MEMS devices can be driven directly or indirectly by piezoelectric energy harvesters. In addition, the working life of piezoelectric energy harvesters is normally larger than that of traditional energy-supply devices. Thus, piezoelectric energy harvesters can improve the service life of MEMS and other electric devices. Moreover, piezoelectric energy harvesters have the merit of a small footprint size. Hence, they have the potential of a high degree of integration with other devices.

In particular, energy harvesters based on human movements are a promising alternative to conventional batteries to solve the issues of a short life span and low energy density. Such harvesters can convert mechanical energy into electric energy to energize low-power electronic devices without affecting human normal motion. Energy harvested from human walking is particularly attractive because of the large amount of harvestable energy and relative minimal effect on the user's comfort compared with other human activities involving limbs. According to practical application, energy harvesters can be classified into three groups based on the situation in which it is worn: direct wearable devices, indirect wearable devices, and indirect unwearable devices.

Various energy harvesting transduction techniques have advantages and disadvantages. To understand the energy output range of each type of transducer under different walking motions, Partridge and Bucknall (Partridge & Bucknall, 2016) conducted an energy flow assessment based on various types of generators. It indicated that the expected energy output for PZT (1.4–3.9 J/step), PVDF (0.2–1.8 J/step), electromagnet (2.0–4.7 J/step), and dielectric elastomers (1.9–5.3 J/step) can serve as potential energy sources to energize electronic equipment with low power consumption. The main objectives in designing energy harvesting devices are to ensure high-power output and self-sustaining capability without affecting the environment.

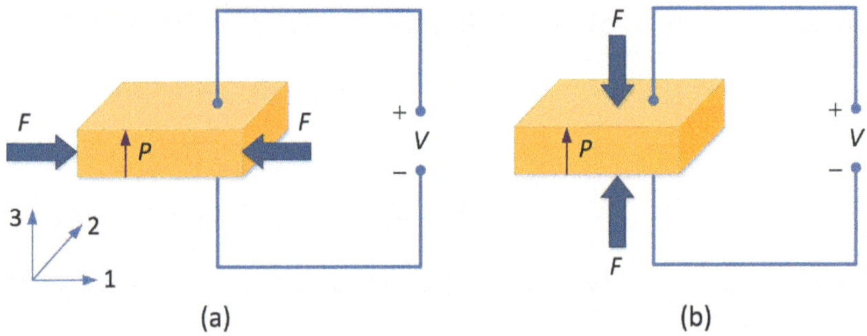

Figure 1.2 Typical working modes of piezoelectric energy harvester, in which F denotes the applied force direction and P indicates the poling direction. (A) 31 mode; (B) 33 mode.

Typically, piezoelectric energy harvesters have two modes of operation: the 31 mode (longitudinal mode) and 33 mode (transverse mode), as illustrated in Fig. 1.2. In the 31 mode, a lateral force is applied in the direction perpendicular to the poling direction. A piezoelectric thin film usually operates in the 31 mode. In the 33 mode, a compressive force is applied in the direction that is the same as the poling direction. A piezoelectric stack generally works in the 33 mode.

1.3 Energy harvesting from human walking

Several piezoelectric energy harvesters have been proposed to scavenge energy excited by human walking (Nia et al., 2017). A review of walking energy harvesting with piezoelectric materials is presented in Nia et al. (2017). According to the form of piezoelectric material used, three kinds of piezoelectric materials are commonly adopted in energy harvesters: piezoelectric patch (Hu et al., 2018; Kang et al., 2012; Somkuwar et al., 2018), piezoelectric disc (Kan et al., 2017; Yang & Zu, 2016), and piezoelectric stack (He et al., 2021; Wang et al., 2021; Zou et al., 2021). Samples of typical piezoelectric materials used in energy harvesting are illustrated in Fig. 1.3.

Based on their characteristics, different piezoelectric materials are used to harvest energy in various situations. Usually, the piezoelectric patch is applied in piezoelectric energy harvesters to scavenge energy excited by a low input force with a particular frequency, because it is easily broken even if the applied force is small. Compared with a piezoelectric patch, the piezoelectric disc can work under large-pressure conditions, and it has a

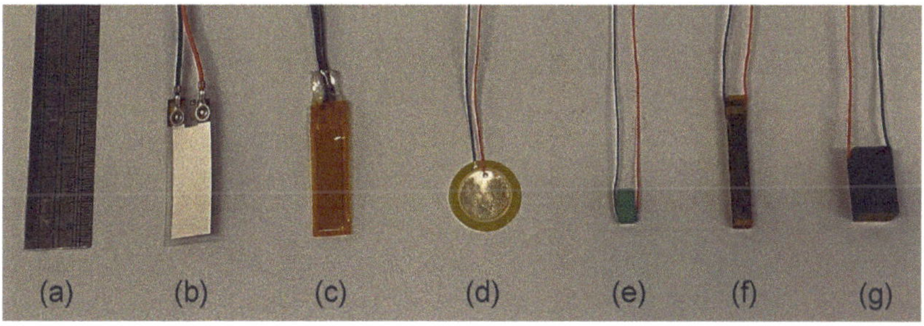

Figure 1.3 Samples of piezoelectric materials in energy harvester. (A) Reference ruler; (B) polyvinylidene fluoride (PVDF) patch; (C) macrofiber composite (MFC) patch; (D) piezoelectric disc; and (E–G) piezoelectric stack.

higher electromechanical transduction rate. The piezoelectric energy harvester excited by a large input force is often constructed with a piezoelectric stack, because it has the highest piezoelectric coefficient and highest stiffness among the three common forms of piezoelectric materials.

1.3.1 Types of piezoelectric energy harvesters

Piezoelectric energy harvesters can be classified into two categories, resonance and nonresonance types. Most piezoelectric energy harvesters are designed based on different cantilever structures with a proof mass fixed at the free end. A piezoelectric film (in the 31 mode) attached to the cantilever can harvest the energy of vibrations experienced by the proof mass. An example is illustrated in Fig. 1.4. This resonance type of piezoelectric

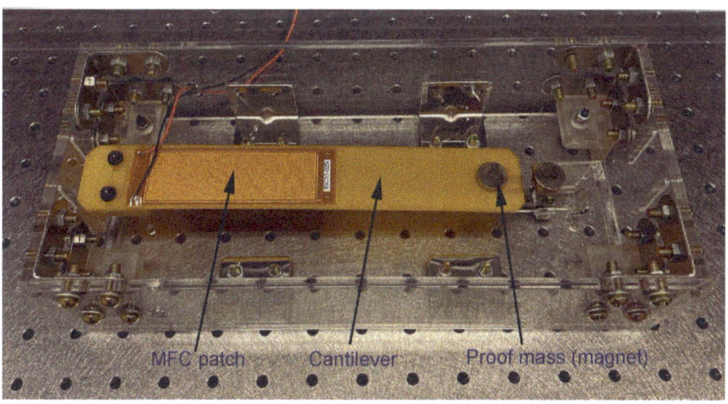

Figure 1.4 A piezoelectric energy harvester with a cantilever beam, a macrofiber composite (MFC) patch, and a proof mass.

energy harvester is able to collect vibration energy from a certain range of excitation frequency. However, its power density is relatively low.

Alternatively, a piezoelectric stack-based piezoelectric energy harvester (in the 33 mode) (Wang et al., 2017) has a higher power density than a piezoelectric energy harvester based on piezoelectric film (in the 31 mode). The piezoelectric stack-based piezoelectric energy harvester can be used as a nonresonance type or resonance type of energy harvester. It has the property of small deflection. It is preferable for harvesting energy from the footsteps of human walking because the frequency of human walking is typically low (around 1 Hz), and a small vertical deflection will not influence walking people too much. An example of a piezoelectric energy harvesting floor is shown in Fig. 1.5.

To harvest energy from the foot strike of human walking, there are two types of piezoelectric stack-based piezoelectric energy harvester: footwear and underfloor. For example, a piezoelectric stack—based small-sized footwear piezoelectric energy harvester was reported in Qian et al. (2018) composed of several force amplifiers and two heel-shaped aluminum plates. An underfloor energy harvesting device was presented in Evans et al. (2019), in which the parameters of the force amplifier and piezoelectric energy harvester were optimized to improve the output performance. Unlike the footwear-type harvester, which is carried by foot, the

Figure 1.5 A piezoelectric energy harvesting floor based on a piezoelectric stack and force transmission mechanism.

underfloor-type harvester cannot be carried away. Instead, it has a larger accommodation space. It can be mounted in crowded areas to improve the frequency of human foot strikes, such as at the entrance of schools, and hospitals.

1.3.2 Force amplification mechanisms

The electric output of a piezoelectric stack is proportional to the load force square acting on the piezoelectric stack (Xu et al., 2013). Thus, to exploit the potential of piezoelectric stack material fully, a force amplification mechanism is required to magnify the force exerted on both sides of the piezoelectric stack to generate large output power. In the literature, displacement amplification mechanisms were widely adopted to amplify small input displacement into large output displacement (Hricko & Havlík, 2020; Kang et al., 2012; Kiziroglou et al., 2020). By swapping the input and output ends of the displacement amplification mechanism, a force amplification mechanism is obtained.

As illustrated in Fig. 1.6, various types of force amplification mechanisms can be applied to magnify the input force (Ling et al., 2016; Liu et al., 2013). Owing to the merits of maintenance free, flexure hinge-based compliant mechanisms have been widely applied in practice (Xu, 2016). In particular, the bridge-type amplification frame (Fig. 1.6B) has the characteristics of a large amplification ratio and compact size, which was previously adopted in the design of a displacement amplifier (Choi et al., 2018; Xu & Li, 2011) and commercialized on the market (e.g., by Cedrat Technologies, Inc.). Nevertheless, a drawback of the traditional bridge-type force amplification frame is that the mechanism has a limited capacity to tolerate the loading force. When different humans walk on the frame (e.g., someone is too heavy or someone is jumping), the force amplification function cannot be properly guaranteed by the loading force exerted on the frame. Hence, the frame must be designed to be as safe as possible to tolerate potential problems.

The safety factor is defined as the ratio between the allowable stress of the frame material and the experienced maximum stress of the harvester. It is used to measure whether the device's material is safe under a loading force. The device is feasible when the safety factor is greater than 1. In previous work, the frame is designed with a thinner and longer beam to ensure a large force amplification ratio (Wang et al., 2017), which leads to a decreased safety factor for the harvester. Researchers have tried to optimize

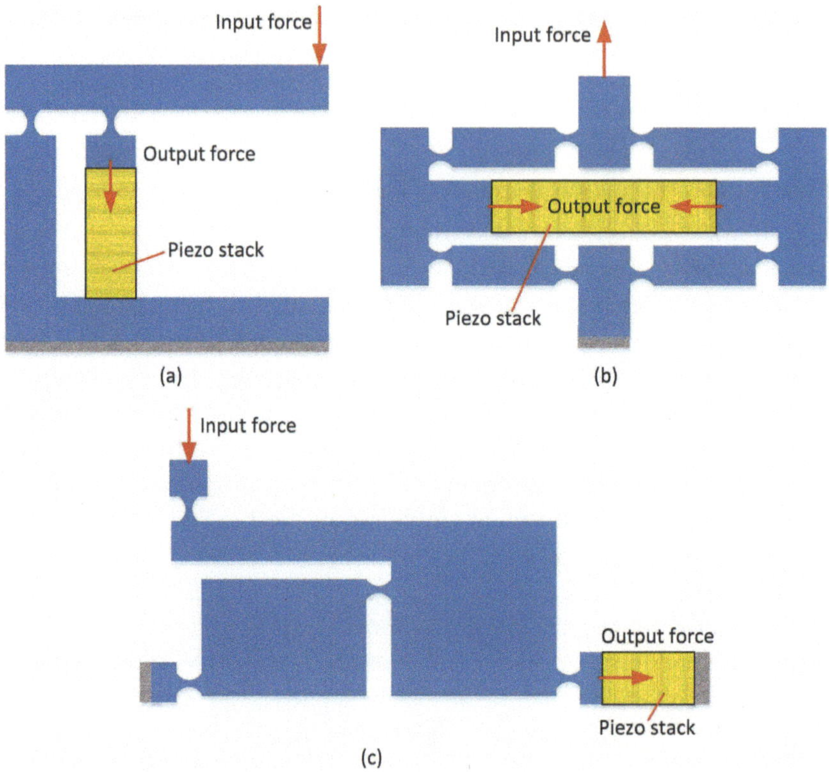

Figure 1.6 Typical force amplification mechanism. (A) Lever mechanism; (B) bridge mechanism; (C) Scott–Russell mechanism.

the force amplification frame by making a trade-off between the safety factor and force amplification ratio. Yet, it is hard to improve the practical behavior of the energy harvester radically.

The output performance of these energy harvesting devices depends on the excitation frequency, mechanical design, and property of the piezoelectric material. The power density, operational bandwidth, and durability are the primary bottlenecks (Li et al., 2011).

1.3.3 Electronic circuit of piezoelectric energy harvester

The circuit representation for a piezoelectric energy harvester is shown in Fig. 1.7. The voltage source V_{PZT} is the open circuit voltage generated by the piezoelectric material. C is the capacitance between two electrodes and R_s is the resistance of the piezoelectric material. In addition, R_L represents an externally connected resistive load.

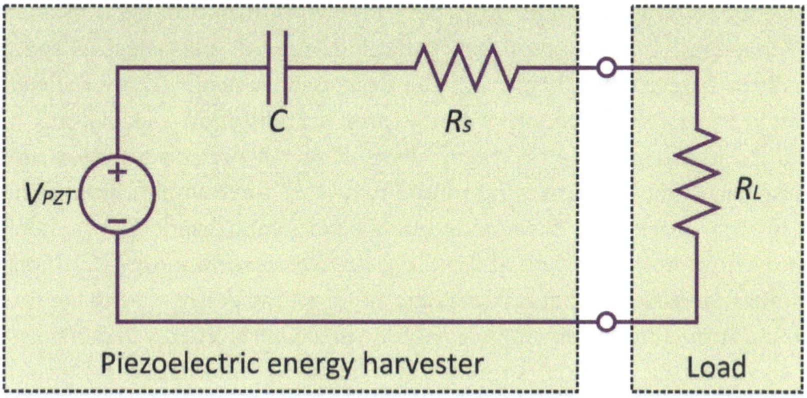

Figure 1.7 Circuit representation of a piezoelectric energy harvester.

For practical use, energy generated by the piezoelectric material is rectified and stored to supply power to external devices, such as MEMS sensors. In the literature, four types of electronic circuits are commonly adopted (Covaci & Gontean, 2020): (1) an AC-DC piezoelectric energy harvesting circuit, (2) a two-stage piezoelectric energy harvesting circuit, (3) synchronized switch harvesters on an inductor circuit, and (4) a synchronous electric charge extraction circuit. These circuits can be adopted in different situations. For example, the first one rectifies AC power generated by the piezoelectric transducers, the second is able to maximize the rectified power, the third adopts a nonlinear technique to improve output power, and the fourth does not rely on the external load.

Moreover, several commercial electronic circuits are available on the market, including MB39C811 from Infineon Technologies AG, LTC3588-1 from Analog Devices, Inc., and MAX17710 from Maxim Integrated, Inc.

1.4 Book outline

This book aims to provide state-of-the-art mechanical designs of various one-dimensional (1-D), 2-D, and 3-D energy harvesters based on a piezoelectric stack. The book discusses how to implement a novel piezoelectric energy harvester to generate electricity from a multidirectional force excited by human walking. It involves analytical modeling, the optimal design, a simulation study, prototype fabrication, and experimental investigations. Detailed examples of analyses and implementations are provided, and perspectives on future development trends are discussed.

A review of energy harvesting devices from human walking is provided in Chapter 2. A survey on mechanical designs of piezoelectric energy harvester is given in Chapter 3. The designs of several 1-D piezoelectric energy harvesters based on various force amplification mechanisms are outlined in Chapters 4—6. Specifically, Chapter 4 presents a piezoelectric energy harvester based on a compound two-stage force amplification frame. Chapter 5 proposes a dual-mode energy harvesting handrail to generate energy from both vibration and pulling force excitation modes. Chapter 6 reports a piezoelectric energy harvester using an integrated multistage force amplification mechanism for scavenging mechanical energy from human footsteps during walking. In addition, Chapter 7 considers a 1-D piezoelectric energy harvester for harvesting energy from bidirectional force excitation. Chapters 8—10 depict the designs of 2-D piezoelectric energy harvesters for scavenging energy from human walking footsteps. In particular, Chapter 8 devises a piezoelectric energy harvester for harvesting energy from two-degrees-of-freedom excitation in vertical directions using a single piezoelectric stack. Chapter 9 develops a 2-D piezoelectric energy harvester with permanent magnets and a multistage force amplification frame. Chapter 10 demonstrates a dual-axial underfloor piezoelectric energy harvester based on a piezoelectric stack. In addition, a 3-D piezoelectric energy harvester is designed and implemented in Chapter 11. Moreover, Chapter 12 outlines the design of a nonlinear bistable energy harvester with enhanced output performance.

References

Choi, K. B., Lee, J. J., Kim, G. H., Lim, H. J., & Kwon, S. G. (2018). Amplification ratio analysis of a bridge-type mechanical amplification mechanism based on a fully compliant model. *Mechanism and Machine Theory, 121*, 355—372. https://doi.org/10.1016/j.mechmachtheory.2017.11.002

Covaci, C., & Gontean, A. (2020). Piezoelectric energy harvesting solutions: A review. *Sensors*, (12).

Du, S., Jia, Y., Zhao, C., Amaratunga, G. A. J., & Seshia, A. A. (2018). A passive design scheme to increase the rectified power of piezoelectric energy harvesters. *IEEE Transactions on Industrial Electronics, 65*(9), 7095—7105. https://doi.org/10.1109/TIE.2018.2798567

Evans, M., Tang, L., Tao, K., & Aw, K. (2019). Design and optimisation of an underfloor energy harvesting system. *Sensors and Actuators, A: Physical, 285*, 613—622. https://doi.org/10.1016/j.sna.2018.12.002

He, L., Zhao, D., Li, W., Wu, X., & Cheng, G. (2021). A dual piezoelectric energy harvester with contact and non-contact driven by inertial wheel. *Mechanical Systems and Signal Processing, 146*.

Hricko, J., & Havlík, Š. (2020). Compliant mechanisms for motion/force amplifiers for robotics. In *Advances in intelligent systems and computing* (Vol. 980, pp. 26–33). Springer Verlag. https://doi.org/10.1007/978-3-030-19648-6_4

Hu, G., Tang, L., Das, R., & Marzocca, P. (2018). A two-degree-of-freedom piezoelectric energy harvester with stoppers for achieving enhanced performance. *International Journal of Mechanical Sciences, 149*, 500–507. https://doi.org/10.1016/j.ijmecsci.2017.07.051

Kan, J., Fu, J., Wang, S., Zhang, Z., Chen, S., & Yang, C. (2017). Study on a piezo-disk energy harvester excited by rotary magnets. *Energy, 122*, 62–69. https://doi.org/10.1016/j.energy.2017.01.059

Kang, Z., Wang, X., & Luo, Z. (2012). Topology optimization for static shape control of piezoelectric plates with penalization on intermediate actuation voltage. *Journal of Mechanical Design, Transactions of the ASME, 134*(5). https://doi.org/10.1115/1.4006527

Kiziroglou, M. E., Temelkuran, B., Yeatman, E. M., & Yang, G. Z. (2020). Micro motion amplification-a review. *IEEE Access, 8*, 64037–64055. https://doi.org/10.1109/ACCESS.2020.2984606

Klein, J., & Zuo, L. (2017). A velocity-amplified electromagnetic energy harvester for small amplitude vibration. *Smart Materials and Structures, 26*(9). https://doi.org/10.1088/1361-665X/aa7e54

Liao, Y., & Liang, J. (2018). Maximum power, optimal load, and impedance analysis of piezoelectric vibration energy harvesters. *Smart Materials and Structures, 27*(7). https://doi.org/10.1088/1361-665X/aaca56

Li, X., Guo, M., & Dong, S. (2011). A flex-compressive-mode piezoelectric transducer for mechanical vibration/strain energy harvesting. *IEEE Transactions on Ultrasonics, Ferroelectrics, and Frequency Control, 58*(4), 698–703. https://doi.org/10.1109/TUFFC.2011.1862

Ling, M., Cao, J., Zeng, M., Lin, J., & Inman, D. J. (2016). Enhanced mathematical modeling of the displacement amplification ratio for piezoelectric compliant mechanisms. *Smart Materials and Structures, 25*(7). https://doi.org/10.1088/0964-1726/25/7/075022

Liu, W. Q., Badel, A., Formosa, F., Wu, Y. P., & Agbossou, A. (2013). Novel piezoelectric bistable oscillator architecture for wideband vibration energy harvesting. *Smart Materials and Structures, 22*(3), 035013. https://doi.org/10.1088/0964-1726/22/3/035013

Mishra, S., Unnikrishnan, L., Nayak, S. K., & Mohanty, S. (2018). Advances in piezoelectric polymer composites for energy harvesting applications: A systematic review. *Macromolecular Materials and Engineering, 304*.

Nellore, K., & Hancke, G. P. (2016). A survey on urban traffic management system using wireless sensor networks. *Sensors, 16*(2). https://doi.org/10.3390/s16020157

Nersesian, R. (2014). *Energy for the 21st century: A comprehensive guide to conventional and alternative sources.*

Nia, E. M., Zawawi, N. A. W. A., & Singh, B. S. M. (2017). A review of walking energy harvesting using piezoelectric materials. *IOP Conference Series: Materials Science and Engineering, 291*, 012026.

Partridge, J. S., & Bucknall, R. W. G. (2016). An analysis of the energy flow and energy potential from human energy harvesting with a focus on walking. *Cogent Engineering, 3*(1). https://doi.org/10.1080/23311916.2016.1215203

Priya, S., Song, H.-C., Zhou, Y., Varghese, R., Chopra, A., Kim, S.-G., & Polcawich, R. G. (2017). A review on piezoelectric energy harvesting: Materials, methods, and circuits. *Energy Harvesting and Systems, 4*(1), 1–34. https://doi.org/10.1515/EHS-2016-0028

Qian, F., Xu, T. B., & Zuo, L. (2018). Design, optimization, modeling and testing of a piezoelectric footwear energy harvester. *Energy Conversion and Management, 171*, 1352–1364. https://doi.org/10.1016/j.enconman.2018.06.069

Somkuwar, R., Chandwani, J., & Deshmukh, R. (2018). Wideband auto-tunable vibration energy harvester using change in centre of gravity. *Microsystem Technologies, 24*(7), 3033–3044. https://doi.org/10.1007/s00542-018-3846-x

Souri, H., Nam, I. W., & Lee, H. K. (2015). A zinc oxide/polyurethane-based generator composite as a self-powered sensor for traffic flow monitoring. *Composite Structures, 134*, 579–586. https://doi.org/10.1016/j.compstruct.2015.08.112

Wang, L., Chen, S., Zhou, W., Xu, T. B., & Zuo, L. (2017). Piezoelectric vibration energy harvester with two-stage force amplification. *Journal of Intelligent Material Systems and Structures, 28*(9), 1175–1187. https://doi.org/10.1177/1045389X16667551

Wang, K., Zhou, J., Ouyang, H., Chang, Y., & Xu, D. (2021). A dual quasi-zero-stiffness sliding-mode triboelectric nanogenerator for harvesting ultralow-low frequency vibration energy. *Mechanical Systems and Signal Processing, 151*. https://doi.org/10.1016/j.ymssp.2020.107368

Xu, Q. (2016). *Design and implementation of large-range compliant micropositioning systems*. Wiley.

Xu, Q., & Li, Y. (2011). Analytical modeling, optimization and testing of a compound bridge-type compliant displacement amplifier. *Mechanism and Machine Theory, 46*(2), 183–200. https://doi.org/10.1016/j.mechmachtheory.2010.09.007

Xu, T. B., Siochi, E. J., Kang, J. H., Zuo, L., Zhou, W., Tang, X., & Jiang, X. (2013). Energy harvesting using a PZT ceramic multilayer stack. *Smart Materials and Structures, 22*(6). https://doi.org/10.1088/0964-1726/22/6/065015

Yang, Z., & Zu, J. (2016). Toward harvesting vibration energy from multiple directions by a nonlinear compressive-mode piezoelectric transducer. *IEEE/ASME Transactions on Mechatronics, 21*(3), 1787–1791. https://doi.org/10.1109/TMECH.2015.2459014

Zhengbao, Y., & Jean, Z. (2016). Comparison of PZN-PT, PMN-PT single crystals and PZT ceramic for vibration energy harvesting. *Energy Conversion and Management*, 321–329. https://doi.org/10.1016/j.enconman.2016.05.085

Zou, D., Liu, G., Rao, Z., Tan, T., Zhang, W., & Liao, W. H. (2021). Design of vibration energy harvesters with customized nonlinear forces. *Mechanical Systems and Signal Processing, 153*. https://doi.org/10.1016/j.ymssp.2020.107526

Further reading

Chen, W., Wang, Y., & Deng, W. (2017). Deformable force amplification frame promoting piezoelectric stack energy harvesting: Parametric model, experiments and energy analysis. *Journal of Intelligent Material Systems and Structures, 28*(7), 827–836. https://doi.org/10.1177/1045389X16657426

Chen, S., Wang, L., Zhou, W., Musgrave, P., Xu, T. B., & Zuo, L. (2015). Optimal design of force magnification frame of a piezoelectric stack energy harvester. In *Proceedings of SPIE - the international society for optical engineering* (Vol. 9435)SPIE. https://doi.org/10.1117/12.2084284

Howell, L. L. (2001). *Compliant mechanisms*. Wiley.

Lobontiu, N. (2002). *Compliant mechanisms: Design of flexure hinges*. CRC Press.

Wang, Y., Chen, W., & Guzman, P. (2016). Piezoelectric stack energy harvesting with a force amplification frame: Modeling and experiment. *Journal of Intelligent Material Systems and Structures, 27*(17), 2324–2332. https://doi.org/10.1177/1045389X16629568

Wen, S., Xu, Q., & Zi, B. (2018). Design of a new piezoelectric energy harvester based on compound two-stage force amplification frame. *IEEE Sensors Journal, 18*(10), 3989–4000. https://doi.org/10.1109/JSEN.2018.2820221

CHAPTER 2

Review of energy harvesting devices from human walking

This chapter reviews work on energy harvesters with a novel mechanical design and high output performance from various sources of walking motion. The functionality, configuration, and characteristics of energy harvesters are investigated with various energy harvesting transduction techniques and material properties. The design requirements of physical size, output performance, working conditions, and structural safety are main challenges for energy harvesters. To address these issues, several potential energy harvesting techniques are demonstrated and for application in real situations.

2.1 Introduction

There has been great interest in extracting renewable power from human movement. As a kind of passive harvesting energy source, energy generation from human activities can be applied directly to power low-consumption electronics without affecting normal human motion. Based on the purpose of the design, it is necessary to select the appropriate human motion from different parts of the body. Common examples include movements of knees, feet, and arms (Fan, Liu, et al., 2017, Fan, Yu, & Tang, 2017; Martin & Li, 2017; Semsudin et al., 2015). According to previous work (Starner & Paradiso, 2004), human walking is particularly attractive because it can produce a significant amount of energy and has a minimal effect on the user's comfort. For example, harvestable energy from human gait is about 13.3 W, considering a human of 68 kg with a walking frequency of two steps per second (Xie & Cai, 2015a,b). On the other hand, Partridge et al. (Partridge et al., 2016) listed the net efficiency for different human activities, such as walking (35−40%), running (45−80%), cycling (25.7%), rowing (19.8%), and arm cranking (23.4%), which showed that walking and running had a higher mechanical conversion efficiency than other activities.

Mechanical Design of Piezoelectric Energy Harvesters
ISBN 978-0-12-823364-1
https://doi.org/10.1016/B978-0-12-823364-1.00005-5
© 2022 Elsevier Inc.
All rights reserved.

The gait cycle includes the stance and swing stages. It is used to measure a complete walking period when the same foot touches and leaves the ground. The right foot is still on the ground when the left foot initially begins to contact the ground during one gait cycle (Fan, Yu, & Tang, 2017; Martin & Li, 2017; Semsudin et al., 2015). Human walking behavior varies with time; it is suitable as a source of power generation. To generate the maximum power output under various gait speeds, the measurement of acceleration for human motion during both running and walking has been studied extensively. For instance, Von Buren (von Buren et al., 2006) and Ylli et al. (Ylli et al., 2015) reported that the maximum power output is produced by the lower leg at a walking speed of 4 km/h. The framework for converting human running from the movements was studied by Elvin et al. (Elvin & Elvin, 2013) and Ylli et al. (Ylli et al., 2015). According to treadmill testing results, the stepping frequency varied between 0.8 and 1.2 Hz under various walking motion speeds of 4 to 10 km/h. Hence, most devices have been designed based on such excitation frequency.

Furthermore, capturing energy from human walking has drawn great attention. Main excitations include the rotary motion of the knee, the weight of a person acting on the shoes or the ground, the accelerating swing motion of the leg, and the acceleration pulse of the heel strike. In view of practical applications, energy harvesters can be classified into three groups based on the wearing status: direct wearable energy harvesters (DWEHs), indirect wearable energy harvesters (IWEHs), and indirect unwearable energy harvesters (IUEHs). Each type of generator has its own features. Several studies have been conducted in the literature.

This chapter provides an overview of energy harvesters based on the excitation source of human walking. Various energy harvesting transduction techniques are demonstrated by exploiting the functionality, configuration, and characteristics. The review also summarizes several potential applicable energy harvesting techniques address the challenges of current transducers.

2.2 Alternative energy sources of batteries

2.2.1 Electromagnetic generators

Electromagnetic generators composed of coil and magnets are relatively mature techniques. They have been developed in every area of life (Klein & Zuo, 2017; Mitcheson et al., 2007; Siddique et al., 2015). The basic principle is that electromotive force is produced by the relative vibration motion of an electric circuit and magnetic field, leading to current flow across the electric load, according to Faraday's law of induction. Electromagnetic transducers with low impedance can produce high current output under various vibration situations. However, output power is limited by low voltage output and a low electromechanical coupling coefficient.

2.2.2 Electrostatic generators

Electrostatic generators can also convert mechanical energy into electric energy. The transduction procedure is operated by the relative movement of isolated capacitor plates. An external voltage source is provided for the initial charging stage. Conversion based on an electrostatic principle differs from other alternative energy sources based on physical characteristics such as the piezoelectric (PZT) and photovoltaic effect. Thus, the control method and mechanical design have a significant role in harvesting more energy (Goudar et al., 2013).

Dielectric elastomer is usually adopted in electrostatic generators. Dielectric elastomer is a kind of rubber-like material with a high energy density. Its energy density is five to 40 times better than that of PZT material (Kornbluh et al., 2012). Thus, electrostatic transducers have the features of high output and great performance for frequency tuning and voltage rectifying (Siddique et al., 2015). However, devices with high output impedance result in low output current and a complex system because of the need for an external voltage supply.

2.2.3 Piezoelectric generators

Electric energy generated from PZT transducers depends on external pressure or vibration. PZT materials based on crystalline structure have positive and negative charges, leading to dipole moments (Li et al., 2014). The distortion of dipoles results in electrical power when they are subjected to mechanical stress. Two major PZT materials are widely employed in energy harvesting: PZT ceramic and polyvinylidene difluoride (PVDF) polymer.

There are several modes of action for PZT material, including d33 and d31 modes. Because the d33 mode produces two or three times power output compared with the d31 mode, it is better to apply PZT material at the d33 mode. However, it is relatively difficult to design a PZT transducer acting in the d33 mode compared with the d31 mode. Typical PZT structures are designed based on a cantilever beam acting in the d31 mode, whereas it is challenging to vibrate PZT beams at resonant frequencies under broadband frequency conditions to achieve maximum power output. Besides, a cymbal configuration could help the d31 mode of a transducer to adjust excitation sources to achieve a high energy output.

2.2.4 Generators with other configurations

In addition, several novel energy harvesting mechanisms have been studied employing other energy sources. For example, airflow-type energy harvesters mainly rely on an external force to form airflow. Electric energy is generated when airflow passes across generators. A triboelectric nanogenerator achieves high electric output based on coupling between electrostatic and triboelectric effects.

Generators based on other configurations can also produce high power output, although they are not popular because of their relative complex structure, high cost, or high requirements for environmental conditions.

2.3 Direct wearable energy harvester

Direct wearable devices have a simple mechanism. They are mounted on the human body to capture vibration energy from the swaying motion of the human gait. A few applications are summarized subsequently.

2.3.1 Electromagnetic devices

Donelan et al. (Donelan et al., 2008) proposed a gear-based knee-joint type of energy harvester. The rotary motion of the knee joint is converted into the motion of gears, which drives the generator to generate electric power. The gear motion is amplified by various sizes of gears. The safety and efficiency of the electromagnetic generator were improved by Chen et al. (Chen et al., 2017). The unique gear transmission mechanism converted the clockwise and counterclockwise rotation from input to a unidirectional rotation of the generator. At a walking speed of 1.5 m/s, the average generated power is 3.6 W, which is better than that of previous single-directional harvesters.

Rao et al. (Rao et al., 2013) designed a complete energy harvesting system containing an electromagnetic generator, interface circuit, and rechargeable battery. The generator is composed of a spherical box, coil, and magnet ball. The electricity was generated by the movement of a magnet ball according to Faraday's law. It harvested energy from various directions of human motion. Moreover, Xie et al. (Xie et al., 2016) proposed an integrated energy harvesting device. The container shell is composed of a spring, microgenerator, and gear rack. External vibration leads to relative motion between the container shell and planetary gear train, which drives the generator to generate power. Wang et al. (Wang et al., 2017) employed a magnet and spring to design tunable electromagnetic generators, as illustrated in Fig. 2.1A. Experimental results show that the device produces electricity under various acceleration levels, which is suitable for a wide range of frequency situation. Performance was enhanced when a high human walking speed was reached. The devices had the ability to generate an average power of 0.53 mW across a resistor of 5 Ω under a walking speed of 9 km/h. Siddique et al. (Siddique et al., 2017) proposed an electromagnetic energy harvester with a T-shaped cantilever beam. Compared with four types of cantilever beam mechanism, configuration D obtained a larger power output by the movement of the coil. An energy output of 3.5 mW at 5.6 Ω was achieved under human movement testing. In view of the advantage of the PZT transducer and electromagnetic transducer, Hamid et al. (Hamid & Yuce, 2017) presented a novel energy harvesting system combining the two transducers to maximize power output, as depicted in Fig. 2.1B. Experimental results demonstrated that the power conversion efficiency increased under several conditioning circuits. The maximum power obtained was 0.55 mW.

2.3.2 Piezoelectric devices

A cyclic translation energy harvester with an up-conversion strategy was developed by Pozzi et al. (Michele & Meiling, 2011), as depicted in Fig. 2.1C, which could produce a high frequency of vibration from several PZT bimorphs by deflecting the mechanical plectrum and the tip of the cantilever beam. By optimizing the parameters of the devices, the obtained maximum power output was about 0.93 mW at 310 Hz under a 20-Ω electric load. Based on a similar idea, Kuang et al. (Kuang et al., 2016) further designed a knee-joint generator by magnetic plucking to eliminate inevitable noise and increase longevity, as shown in Fig. 2.1D. Generators with eight PZT bimorphs and 32 magnets had the ability to acquire an average power of 5.8 mW at 0.9 Hz. In addition, the work applied the energy harvester with a mechanical plucking technique to power a wireless sensor to characterize the transducer (Kuang & Zhu, 2016).

Owing to gravity, vibration energy from a swaying motion is different with time during walking. Cao et al. (Cao, Wang, et al., 2015, Cao, Zhou, et al., 2015) constructed a nonlinear time-varying bistable energy harvester to improve the performance and effectiveness of the devices. The

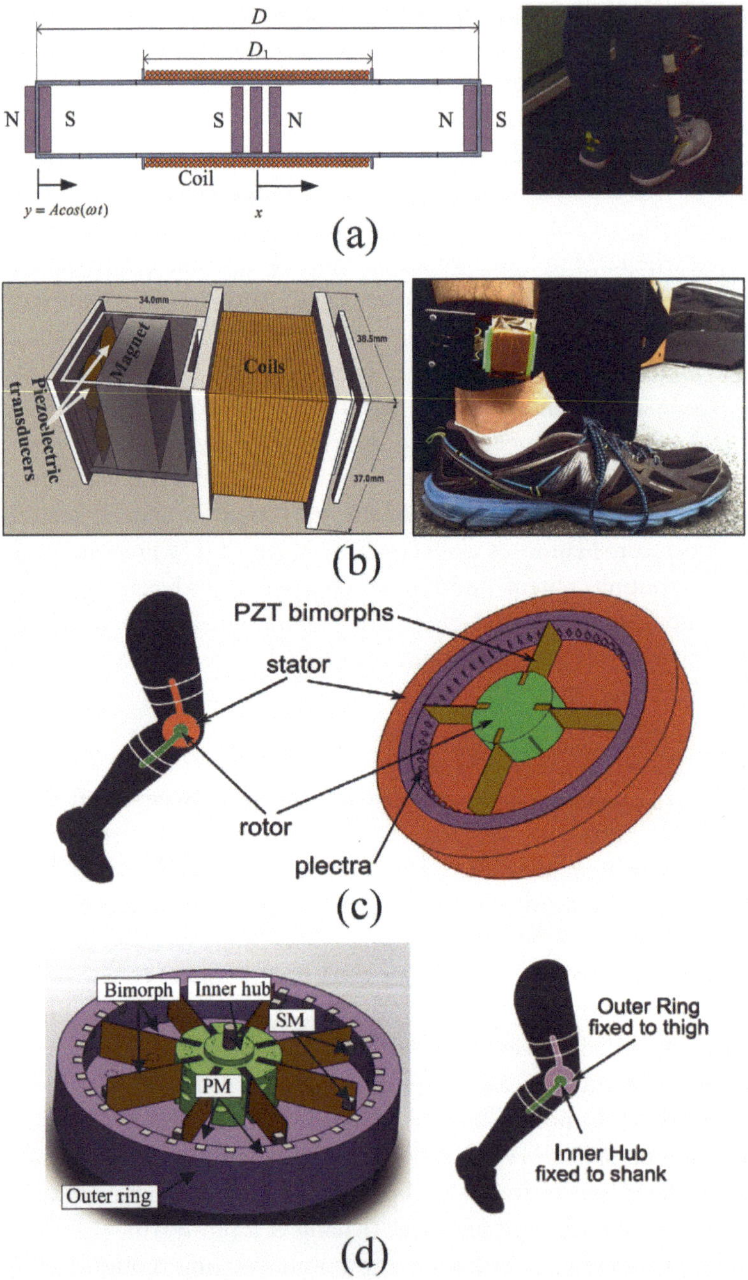

Figure 2.1 Illustrations of energy harvesting from human walking. (A) Magnetic-spring electromagnetic generator (Wang et al., 2017); (B) hybrid generator (Hamid & Yuce, 2017); (C) mechanical plucking type of generator (Michele & Meiling, 2011); (D) magnetic plucking type of generator (Kuang et al., 2016).

maximum energy output was approximately 0.0146 mW at a walking speed of 8 km/h. In addition, Wang et al. (Wang et al., 2015) employed a tristable configuration to improve the power output under similar testing conditions and mechanical design.

2.3.3 Summary

Direct wearable devices offer a convenient way to capture vibration energy from the environment. However, the frequencies of human activities are low, up to a few Hertz. To improving the energy output, devices with the function of up-conversion frequencies are effective, so the devices are able to vibrate at resonance under low frequency of external excitation. Most researchers tend to employ this idea using various approaches (Halim & Park, 2014a, 2014b; Kwon et al., 2017; Michele & Meiling, 2011). The basic theory is that vibration energy from the environment with low frequency input is transferred to another system by mechanical design, leading to a second vibration with a high frequency output. The drawbacks are the weight and mechanical complexity. Moreover, DWEH offers worse user comfort because of long-term use. In addition, to enhance the performance of the energy harvester, various types of nonlinear cantilever mechanisms, such as monostable (Zhao & Erturk, 2013; Zhou et al., 2016), bistable (Cao, Wang, et al., 2015, Cao, Zhou, et al., 2015; Harne & Wang, 2013; Kim, Hwang, et al., 2016, Kim, Nguyen, et al., 2016, Kim, Son, & Seok, 2016), and tristable (Cao, Wang, et al., 2015) generators, are actively investigated to adapt to a wide range of frequency conditions.

2.4 Indirect wearable energy harvester

Indirect wearable devices are usually mounted on wearable items, such as backpacks and shoes, to scavenge vibration energy indirectly from human walking. They are able to eliminate the impact of the inconvenience and inflexibility of DWEHs. Such devices with a compact size and light weight increase portability and utility, but they are prone to reduce output power because of the physical size restriction. Two major types of design are described next.

2.4.1 Shoe-mounted devices

The most popular application is to embed the energy harvester in a shoe to capture energy from periodic foot strikes between the ground and heel. The space in which to place the energy harvester in a shoe is limited.

Kymissis et al. (Kymissis et al., 1998) derived the development, characteristics, and potential application of PZT generators and magnetic generators. The authors suggested several approaches for enhancing the power output in view of the number of material and mechanical characteristics.

2.4.1.1 Electromagnetic devices

Purwadi et al. (Purwadi et al., 2015) studied a gear-based generator that harvested energy from foot strikes. The linear motion of walking force was converted into the rotary motion of gears, which drove a small-scale generator to generate electricity. Xie and Cai (Xie & Cai, 2015a,b) further developed an in-shoe electromagnetic energy harvester with an amplification mechanism based on a trapezoidal slider configuration to manipulate the displacement input, as shown in Fig. 2.2A. An up-conversion frequency structure of gear train was studied to maximize the power output. During human walking, the harvester was driven by foot strikes at the stance phase. When the generator was released at the wing phase, the restoring energy in the springs would continue to press the generator to generate electricity. Ylli et al. (Ylli et al., 2015) studied two different multicoil topology harvesters based on the characteristics of human footsteps. The first is the swing-type generator, as illustrated in Fig. 2.2B, which mainly harnesses mechanical energy from the swing phase during human walking. It varies from other shoe-mounted generators based on the heel strike. An average power output of 0.84 mW is obtained. The second is a typical shock-excited energy harvester based on a magnetic mechanism, as shown in Fig. 2.2C, which can generate a power of 4.3 mW. Rahman et al. (Rahman et al., 2015) presented an electromagnetic generator based on microfluidics, which was mounted on shoe insoles to harness energy. By optimizing parameters and spans of coil, the performance of the transducer was tested under various frequencies, which showed that a total power of 455.77 mW was achieved. Liu et al. (Yulong et al., 2016) designed a shoe-mounted transducer based on a magnetic gear and permeant magnet, as depicted in Fig. 2.2D; it solved the mechanical issue caused by the transmission mechanism. The maximum power output was 1.4 W at a total rotating time of 1 s. Wu et al. (Wu et al., 2017) studied a compact three-degrees-of-freedom (3-DOF) electromagnetic generator. It included a permanent magnet, two elastic strings, and ferrofluid. It harvested more vibration energy from human walking and minimized the influence of frictional loss.

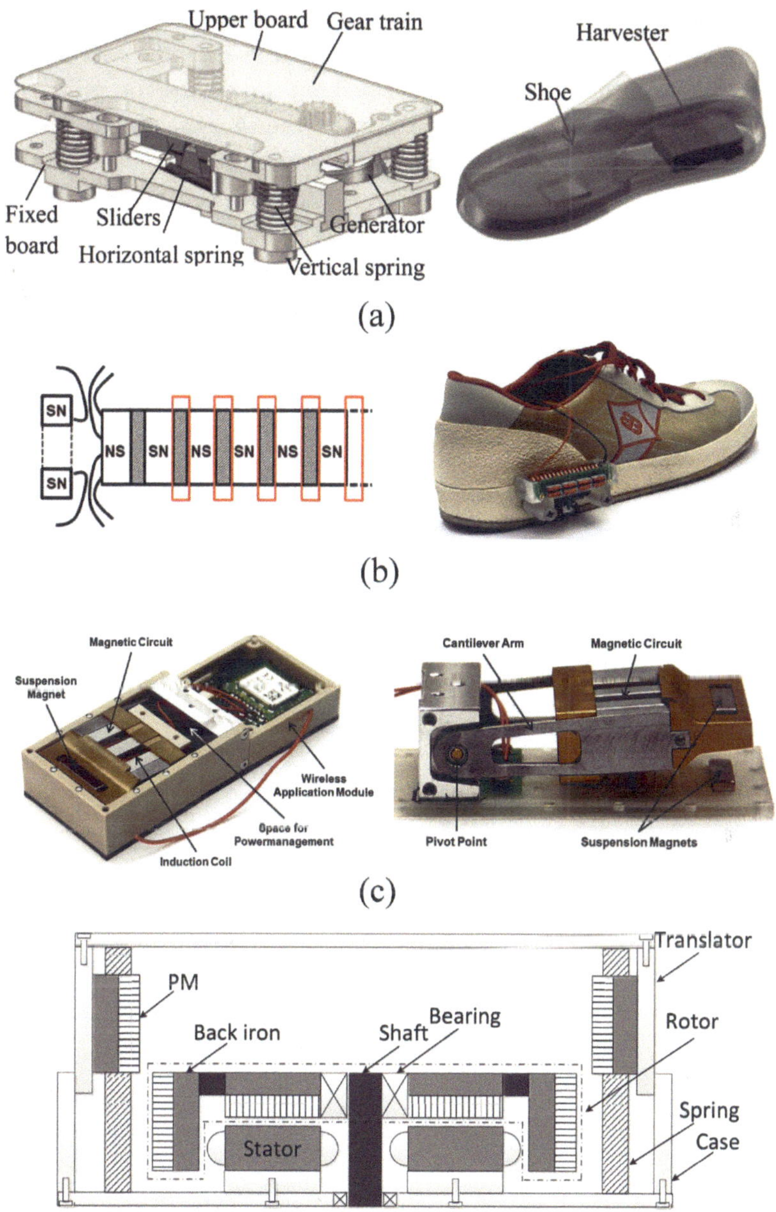

Figure 2.2 Illustrations of shoe-mounted devices. (A) Electromagnetic energy harvester based on trapezoidal slider configuration (Xie & Cai, 2015a,b); (B) swing-type electromagnetic energy harvester (Ylli et al., 2015); (C) shock-excited type electromagnetic energy harvester (Ylli et al., 2015); (D) shoe-mounted transducer based on magnetic gear and permeant magnet (Yulong et al., 2016).

2.4.1.2 Piezoelectric devices

Howell (Howells, 2009) demonstrated a heel-strike unit and conducted an evaluation for practical application based on suitability for military, as shown in Fig. 2.3A. A sensitivity analysis of the shoe-mounted generator was developed by Moro et al. (Moro & Benasciutti, 2010), which improved energy conversion efficiency under imposed constraints. This device mainly captured vibration energy from swing motion. The power output was restricted at the microwatts level because of the compact size requirement for in-shoe devices and low frequency excitation from footsteps. Moreover, the amount of PZT material used in the devices was not enough to provide a large power output. Concerning the compact size requirements for most shoe-mounted devices, Zhao et al. (Zhao & Erturk, 2013; Zhou et al., 2016) designed a PZT energy harvester with a sandwich structure. Multilayer PVDF film with good flexibility and stability was used to provide comfort to the wearers, as depicted in Fig. 2.3B. The device integrated a power management circuit to obtain high output performance; it

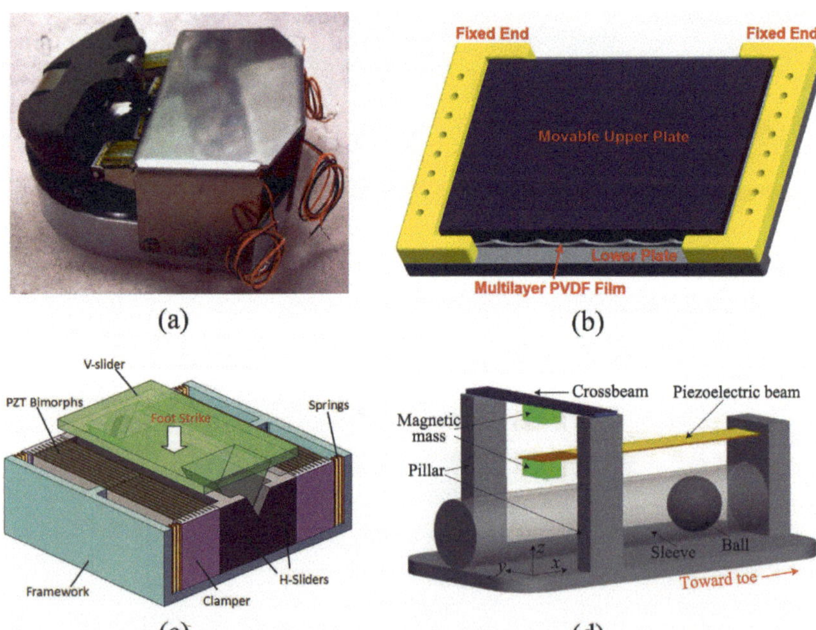

Figure 2.3 Illustration of piezoelectric devices. (A) Heel strike piezoelectric energy harvester (Howells, 2009); (B) schematic of energy harvester with sandwich structure (Zhao & You, 2014); (C) V-shaped energy harvester (Xie & Cai, 2014); (D) piezoelectric energy harvester composed of U-shaped structure (Fan, Liu, et al., 2017).

produced an average power output of 1 mW at a walking frequency of 1 Hz. Ishida et al. (Ishida et al., 2013) developed an in-shoe energy generator based on eight-layer stacks of PVDF sheet. It reached an average power output of 1.1 mW at a walking frequency of 1 Hz. An PZT energy harvester embedded in a shoe to harness the impact energy of heel strikes was proposed by Xie and Cai (Xie & Cai, 2014), as illustrated in Fig. 2.3C. The harvester is composed of a V-shaped amplification mechanism and PZT bimorphs. It produced a maximum power of about 28 mW at a walking frequency of 1.5 Hz. Paul et al. (Paul et al., 2015) studied a PZT generator combined with wireless communication techniques. It was able to connect with a smartphone by a wireless network to achieve more functions. Saha et al. (Saha et al., 2014) simulated and modeled a shoe-embedded PZT energy harvester. The energy output was about 8.4 mW across a resistive load of 500 kΩ. Fan et al. (Fan, Liu, et al., 2017) presented a shoe-mounted PZT harvester to scavenge vibration energy during walking, as shown in Fig. 2.3D. The harvester was composed of a U-shaped structure, magnetic mass sleeve, ball, and PZT beam. It was able to capture energy from three excitation sources with various directions. A U-shaped structure was applied to bear compressive force from the heel. Energy from the swing motion of the leg was harvested by the motion of a ball in the sleeve. A PZT beam was employed to capture vibration energy from the z-direction.

2.4.1.3 Devices with other configuration

Fu et al. (Fu et al., 2015) presented an in-shoe energy harvester based on an air-pumped turbine system to deal with the problems of a low excitation frequency and small input stroke. The device employed an amplification mechanism of air speed by changing the sectional area of the inlet. The performance was tested by normal footsteps from a 65-kg person, which indicated that 6 mW power was achieved at a 4.9-Ω load. The drawback of this structure was low energy conversion efficiency. A bladder-turbine energy harvester with a simple structure and light weight based on airflow and a DC generator was further developed by Fu et al. (Fu et al., 2016). Vertical linear motion from shoes was converted into airflow, leading to rotational movement of the turbine. Energy was generated by the electric generator. The maximum energy output of a 65-kg person was 90.6 mW with a resistance load of 30.4 Ω. Employing a power management circuit, a power of 14.8 mJ was obtained from 165 footsteps.

2.4.2 Backpack-mounted devices

There are also a few generators mounted on wearable items such as backpacks. For instance, Kuo (Kuo, 2005) analyzed the applicability of a backpack-mounted energy harvester from the economical aspect of human walking. The author concluded that such devices could save metabolic energy and produce useful electric power.

2.4.2.1 Electromagnetic devices

Rome et al. (Rome et al., 2005) designed a backpack-mounted energy harvester with a gear chain. Electricity was generated by moving parts of a rack gear system that drove the motion of the generator. Xie et al. (Xie & Cai, 2015a) further studied this type of energy harvester and increased its working efficiency. Although this device could harness the electric power of 4.1 W under a 15-kg load at a walking velocity of 1.56 m/s, the size and weight were still serious problems for practical use.

2.4.2.2 Piezoelectric devices

Feenstra et al. (Feenstra et al., 2008) proposed a PZT energy harvester with a force amplification frame to capture energy from a backpack. Berdy et al. (Berdy et al., 2015) proposed an electromagnetic magnetic generator to generate the power output of 342 μW at a walking speed of 6 km/h, as illustrated in Fig. 2.4A. The device was less sensitive to the swing motion of people, and resulted in decreased power output. Granstrom et al. (Granstrom et al., 2007) studied a backpack-mounted energy harvester with PZT straps. It is a feasible way to collect energy from human walking because of minimal side effects that disrupt the user's comfort. Zhao et al. (Zhao et al., 2016) further designed devices as depicted in Fig. 2.4B. The generator harnesses mechanical energy from various directions of excitation under triple operating frequencies owing to the flexibility and stretchable of a spiral beam. A maximum power output of 330 μW at 16 Hz was generated by bending the beam.

2.4.2.3 Devices with other configuration

In addition, Yang et al. (Yang et al., 2013) developed a self-powered backpack based on triboelectric nanogenerator. The energy conversion efficiency reached up to 10.62%. The device was able to light 40 commercial LED lights under normal human walking.

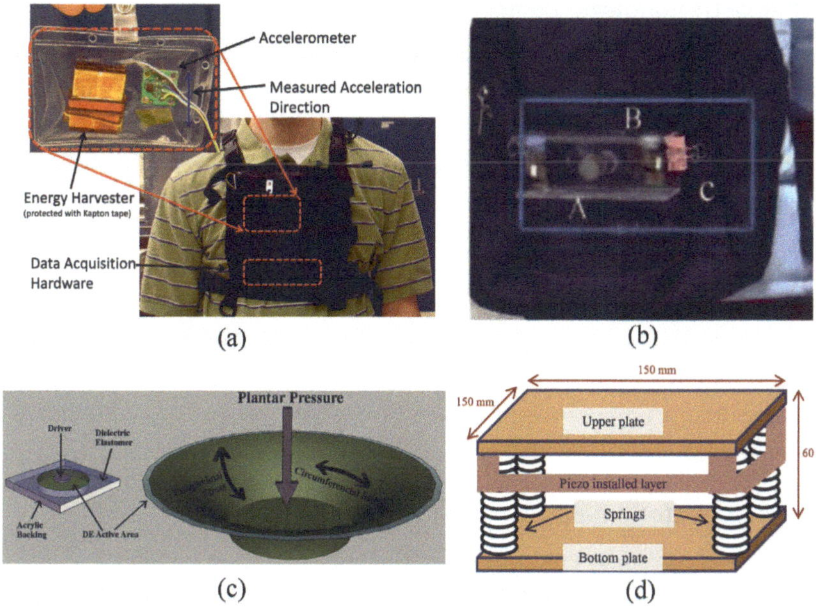

Figure 2.4 Illustrations of (A) electromagnetic magnetic generator (Berdy et al., 2015); (B) piezoelectric generator (Zhao et al., 2016) mounted on the backpack; (C) dielectric elastomer (DE)-based energy harvester (Goudar et al., 2013); (D) mechanical design of piezoelectric tile (Hwang et al., 2015).

2.4.3 Summary

The generator mounted on a backpack could capture limited power at the microwatt level owing to the low working frequency of the backpack. By contrast, shoe-mounted energy harvesters receive several excitations from human walking, which theoretically can generate more power to improve structural issues. Thus, researchers attempt to develop shoe-embedded energy harvesters with high output performance, although the space to place the harvester in a shoe is small. It is a critical challenge to design a compact device with more functions. An effective approach is to find a balance between high power output and small-scale mechanical configuration. A more serious problem is that wearable items are extra components, so the frequency of use and inevitable size and weight are still undesirable according to human behavior.

2.5 Indirect unwearable energy harvester

Indirect unwearable devices harness mechanical energy from compressive pressure such as human weight. They are usually fixed under the ground to energize low-power electronic devices.

2.5.1 Electromagnetic devices

In early studies, Starner (Starner, 1996) proposed an electromagnetic generator with a rotary mechanism that could produce a power output of 0.23 W in 3-cm displacement input. A mesoscale permanent magnetic linear electromagnetic generator with high electric output and economical use was developed by Zeng et al. (Zeng & Khaligh, 2013). A maximum power output of 497.7 mW was achieved by 165 turns of coil windings with an input stroke of 26.7 mm. These configurations were unrealistic because of a large input stroke. To ensure a normal walking experience, the displacement input should not be larger than 10 mm. Sekhar et al. (Sekhar et al., 2014) demonstrated an electromagnetic power generation method and suggested a framework of design progress to provide a cost-effective generators.

2.5.2 Electrostatic devices

Goudar et al. (Goudar et al., 2013) presented a DE-based energy harvester with a novel adaptive closed-loop control algorithm to improve the conversion efficiency and maximize power output, according to the optimal prediction of human gait between footsteps, as illustrated in Fig. 2.4C. The performance was verified by analytical modeling and experiments, which showed that the devices could yield a maximum power of 120 mJ per footstep.

2.5.3 Piezoelectric devices

A typical PZT energy harvester is based on a PZT unimorph, bimorph, or PVDF mounted on a cantilever beam. Strain energy is harnessed by deflecting the beam. Thus, the devices are unable to bear a large force unless a protection mechanism is attached to ensure structural safety. These issues have been investigated. Hwang et al. (Hwang et al., 2015) designed a PZT tile based on a cantilever energy harvesting mechanism, as illustrated in Fig. 2.4D. The device included springs and a supporting plate. It could bear a high load when it was pressed within a certain displacement, until the springs were fully compressed. The extracted power from the devices was 53.1 mW at a matching impedance of 15 kΩ under a testing subject of 68 kg. Kim et al. (Cao, Wang, et al., 2015; Harne & Wang, 2013; Kim, Nguyen, et al., 2016) proposed a footstep-based energy harvester composed of gears, spring, and a revolving mechanism. The compressive force exerted by the human gait was converted to a rotary motion of gears, resulting in the vibration of four cantilever beams. Connections between the bottom gear and cantilever beam was a kind of up-conversation frequency mechanism that could enhance the output performance. Chua et al. (Chua et al., 2016) presented a PZT tile-type energy harvester. It obtained a maximum voltage of 5.56 V for a walking motion, but the structure was relatively simple and lacked an appropriate protection mechanism to ensure the safety of the structure, even though rubber foam was applied in the design. Choo et al. (Choo et al., 2017) demonstrated a novel tuned mass piezo-damper by replacing the damping component of a typical tuned mass damper with a PZT element, so the device became an energy harvester to capture ambient vibration energy, as shown in Fig. 2.5A. Moreover, it had the advantage of a conventional one, such as bearing high force and reducing the acceleration impact on the structure.

Although these energy harvesting devices have the ability to acquire a high energy output, they have room for further improvement, because the space of the devices was not fully exploited. Moreover, the height of the

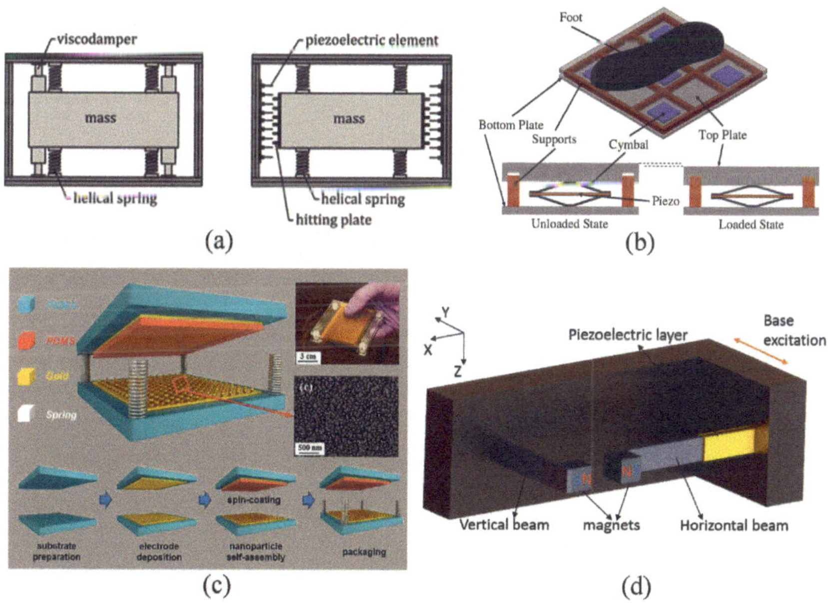

Figure 2.5 Illustrations of (A) typical tuned mass damper (left) and tuned mass piezo-damper (right) (Choo et al., 2017); (B) floor tile energy harvester (Nathan et al., 2016); (C) triboelectric nanogenerator based on coupling between electrostatic effect and triboelectric effect (Zhu et al., 2013); (D) nonlinear energy harvester with internal resonance (Yang & Towfighian, 2017).

transducers should be limited, especially for tile-type generators. Sharpes et al. (Nathan et al., 2016) further developed a floor tile energy harvester based on a PZT cymbal structure to save more space, as illustrated in Fig. 2.5B. There were potential structural safety issues owing to the breakable PZT bimorph when the PZT beams operated in a tension model such as the cymbal structure. To address this challenge, the device had position protection and a limitation mechanism, which also minimized the device deflection's impact on humans while keeping the harvester safe. Furthermore, the device included a wireless commutation technique with the same dimension as typical standard size tiles, which could be regarded as a self-powered switch to control the equipment in a smart building application. An exploratory study on an in-shoe energy harvester based on a textile coil inductor was introduced by Cho et al. (Cho et al., 2016). Although the harvested power was limited at the microwatt level, it provided a novel possibility for mechanical energy harvesting. The flexibility and compactness of the device produced a wide range of room for further improvement.

2.5.4 Devices with other configuration

A novel triboelectric nanogenerator based on coupling between electrostatic and triboelectric effects was presented by Zhu et al. (Zhu et al., 2013,) to achieve high electric output, as shown in Fig. 2.5C. It could generate energy output up to 1.2 W to light at least 600 multicolor LED bulbs arranged in series. Later, Zi et al. (Zi et al., 2016) demonstrated that the triboelectric nanogenerator with a potential voltage of 10–100 V had better output performance than an electromagnetic generator (0.2–4V), which was well-adapted to power small electronics owing to the features of low cost, light weight, and high energy output. Using the concept of a smart floor, He et al. (He et al., 2017) reported a triboelectric nanogenerator to harness vibrational energy without influencing the quality of the floor by employing an aluminum ball and polytetrafluoroethylene film. The device produce 238 V with 2.4 μA.

2.5.5 Summary

Because of their methods of use, because IUEHs do not come into contact with people, they offer the best user comfort. Moreover, they can be regarded as smart self-powered sensors to control relative building systems, such as lighting and climate controls for the fast-emerging IoT (Alsharif et al., 2019; Kamruzzaman et al., 2017; Shaikh & Zeadally, 2016; Valenta & Durgin, 2014). These types of energy harvesters are suitable only for crowded areas such as schools, restaurants, and shopping malls.

2.6 Potential applicable energy harvesting techniques

A few energy harvesting techniques have not been employed in the practical application of human walking because of immature conditions such as the restriction of size and resonant vibration at low walking frequency. It is preferable to explore some potential applicable energy harvesting techniques that are possible to be used in human walking motion and to improve the corresponding working performance.

2.6.1 Direct wearable energy harvester

For DWEH devices, the rotary mechanism is an essential part of most generators to harvest mechanical energy from knee joint motion, and it has attracted significant research interest. For example, Dai (Dai, 2016) developed a vibration energy harvester with a rotary mechanism. The device is composed of six magnets, a rotary pendulum, and magnetoelectric transducer with broadband frequency. It produced an power output of 970.2 μW at a resonant frequency of 14.8 Hz. Fu et al. (Fu & Yeatman, 2017) investigated a PZT energy harvester with a frequency up-conversion configuration to obtain low-speed broadband performance. The frequency conversion was conducted by magnetic plucking, generating more vibration energy from the PZT beam. Kan et al. (Kan et al., 2017) developed a piezo-disk energy harvester based on rotary magnets and showed that the number of magnets had a large impact on the rotary speed and frequency bandwidth.

2.6.2 Indirect wearable energy harvester

Concerning IWEH devices, an effective approach to improving the frequency bandwidth is to optimize current energy harvesters with more functionality and feasibility. For example, Huang and Tsai (Huang & Tsai, 2016) presented an adjustable PZT energy harvester based on conventional cantilever beam structures. The frequency range reached ±30%. However, to maximize the output performance, it is best to design devices with multiple DOF to take full advantage of three excitation sources from human walking.

Hu et al. (Hu et al., 2018) demonstrated a 2-DOF PZT generator with a stopper to enhance performance. With the benefit of the stoppers, the mechanism exhibited various dynamic characteristics that extended the frequency range for broadband use. Experimental results showed that the maximum power output was 7.66 mW across a 50-kΩ resistance load. Yang et al. (Yang & Towfighian, 2017) developed a nonlinear energy harvester with internal resonance, as depicted in Fig. 2.5D. The device was composed of a cantilever beam, tip, PZT bimorph, and permanent magnet. It could be applied to a wide range of frequencies because the 2-DOF vibration system broadened the frequency bandwidth with 97% improvement according to the internal resonance. The maximum energy output was 110 μW at an optimal resistance load of 600 kΩ.

2.6.3 Indirect unwearable energy harvester

For IUEH devices, it is suggested to develop a tile-type transducer able to be applied under high load. The use of a PZT stack is such an attempt (Zhao & Erturk, 2014). Because a PZT stack contains multiple layers of PZT bimorph that can sustain a large amount of pressure, it is well-suited for application in a footstep-based energy harvester to convert a compressive force such as human weight into electricity. However, harnessing vibration energy is not suggested, because its high stiffness leads to a high resonance frequency.

A few studies on stack-based energy harvesters have been conducted. Li et al. (Li et al., 2014) demonstrated a PZT transducer based on a flex-compressive mechanism. Because both ends of the transducer were fixed, the ring-type PZT stack was pressed by the force amplification mechanism, resulting in the generation of electric energy. Experimental result indicated that the generating maximum power output is 14.6 mW across 40 kΩ electric load under the excitation of 1 g at 87 Hz. A study on PZT stack energy harvesting was conducted by Xu et al. (Xu et al., 2013). The authors found that the generated power was much higher than for the cantilever-type harvester under a similar weight and size. Moreover, they mentioned that the energy output had a linear relationship with the force input and strongly depended on dynamic pressure. Wang et al. (Wang et al., 2017) designed a PZT energy harvesting device based on a flex-compressive mode, as depicted in Fig. 2.6A. The device was able to bear a high load owing to the structural mechanism and could obtain electric energy from two PZT stacks. The generated power was 17.8 mW across 120 kΩ under a compressive force of 600 N.

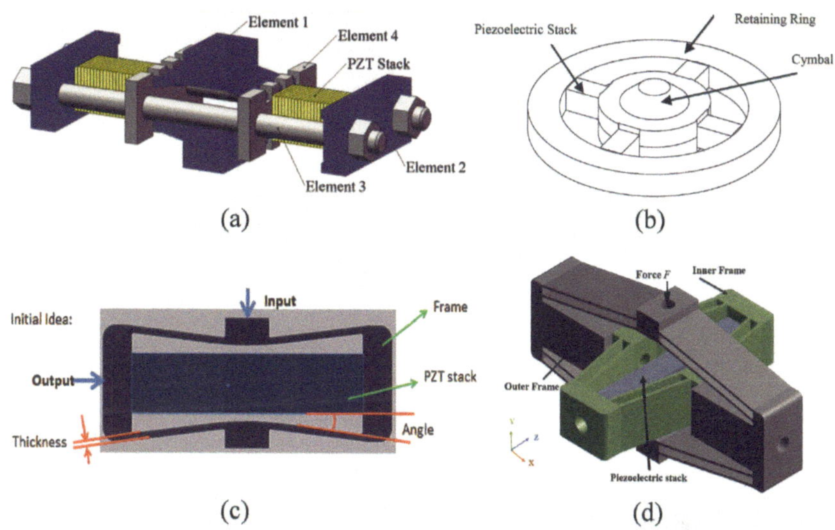

Figure 2.6 Illustrations of (A) piezoelectric energy harvesting device based on flex-compressive mode (Wang et al., 2016); (B) stack-based generators with a circular cymbal structure (Purviance et al., 2013); (C) piezoelectric stack generator with amplification frame (Chen et al., 2015); (D) piezoelectric energy harvester with compound two-stage force amplification frame (Wen et al., 2018).

Purviance et al. (Purviance et al., 2013) developed stack-based generators with a circular cymbal structure, as illustrated in Fig. 2.6B. The device contained four PZT stacks to obtain more energy owing to the benefit of the improved cymbal design. The compressive force was shared by four PZT stacks, resulting in a large force input. Chen et al. (Chen et al., 2017) investigated a PZT stack generator with an amplification frame to magnify the applied force to increase the power output, as illustrated in Fig. 2.6C. The effect of force magnification was equal to the reduction in displacement according to energy conservation, which meant that smaller displacement could produce a larger force output under the same energy. To achieve a high force amplification ratio to generate high power output, various parameter optimization methods and mechanical designs of PZT stack transducers were developed based on a force magnification frame.

To enhance the force amplification ratio within the limited space and reduce the resonance frequency, Wang et al. (Lirong et al., 2016) proposed a two-stage force amplification-based PZT energy harvester, which obtained an 18% energy transmission ratio and 21 times force magnification. Wen et al. (Wen & Xu, 2018) improved the two-stage force amplification

Table 2.1 Output performance of various electromagnetic generators.

Type	Size: L × W × H (mm^3)	Excitation	Power (mW)	Resistor (kΩ)	Reference
SM	80 × 47 × 22	Speed varied from 3 to 5 km/h	810–1390	–	
SM	60 × 40 × 20	Personal weight	4.13	0.166	
SM	60 × 60 × 60	Total 1 s rotating time; 0.1 s for footfall; 0.1 s for foot lifting	1400	–	
DM	70 × 7 × 12.5	Personal weight	0.84	0.166	
DM	60 × 40.7 × 60	Walking speed at 9 km/h	0.53	0.005	
DM	90 × 62 × 28	Walking speed at 0.8 m/s^2	0.22	0.005	

DW, direct wearable; *SM*, shoe-mounted.

frame with high output performance by conducting multiparameter optimization. According to the analysis of the amplification frame and beam, Wen et al. (Wen et al., 2018) further designed a PZT energy harvester with a compound two-stage force amplification frame to increase the safety factor and achieve high power output, as shown in Fig. 2.6D.

These stack-based energy harvesters are immature for practical application because of their lack of a position protection mechanism to ensure structural safety in case the pressing force exceeds the mechanical design limit. The properties of stack-based energy harvesters are listed in Table 2.1. Because the devices have the features of compact size, high power output, high capacity of bearing a large force, and a low displacement input, there is room for further developing such energy harvesters.

2.7 Further discussion

Concerning energy harvesting transduction techniques, the most mature technique is electromagnetic harvesters because of the low cost, low internal impedance, and simple configuration. However, minimizing the physical size to the targeted device domain is a challenge posed to researchers. Moreover, because most of energy harvesters require large deflection, the mechanical design is important. Because of the high energy density of the material, the use of electrostatic generators is an effective solution, whereas the mechanical design of moving components is a threat to practical application. Furthermore, the extra voltage source is a burden on the design. These factors lead to the fact that fewer generators have been developed. Finally, PZT transducers with high energy density have a significant role in vibrational energy harvesting. Compared with other mechanisms, the size and structure of the generators can be designed to become more compact and simpler owing to high electromechanical coupling and easy mechanical energy conversion. The drawback is that the PZT material is too thin, and it cannot bear much force. Thus, it is not easy to justify which type of transducer is the best. It depends on the specific architectural purpose, practical requirements, and proper mechanical design.

In view of the output performance, Tables 2.1 and 2.2 list electromagnetic energy harvesters and PZT energy harvesters. Electromagnetic devices tend to produce larger energy output, but their physical sizes are larger than PZT generators.

According to Table 2.2, for PZT transducers, the maximum power output of a mechanical design depends on the number and properties of the PZT material used in the design. Direct wearable devices produced less electric output than the others owing to less vibration excitation sources in the walking swing phase. In view of the restriction of size, energy generated from shoe-mounted energy harvester is limited. However, concerning the frequency of use, indirect wearable devices have more advantages and more potential to produce more energy output. In addition, indirect unwearable generators produce more power output in the experiment, but the structural size is a critical problem, especially the height of the devices.

To design an applicable generator dedicated to human walking motion, some key factors should be considered:

1. Low stepping frequency is distinctive aspect of walking motion. An up-conversion mechanism, nonresonant mechanism, or amplification mechanism is essential. They can solve issues for a harvester that cannot vibrate at resonance at a higher frequency.

Table 2.2 Output performance of various piezoelectric generators.

Type	Material	Size: L × W × H (mm^3)	Excitation	Power (mW)	Resistor (kΩ)	Reference
DW	PZT-5H	—	Motor	0.93	20	
DW	PZT-51	—	63-kg person walking at 6—9 km/h	0.0146	100	
DW	PZT-51	—	Walking at 4—9 km/h	0.0164	—	
DW	PZT-5H	—	Motor	5.8	15	
IW	PZT-5H	45 × 30 x 24	65-kg person walking at 2—8 km/h	0.03—0.35	120	
IW	PZT-5H	50 × 40 × 23	68-kg person walking at 1—1.5 Hz	28	0.15	
IW	PZT-5A	—	Shaker	8.4	0.5	
IW	PZT, PVDF	80 × 50 × 3	Walking at roughly 1 Hz	1	1268	
IW	PZT	33.8 × 20 × 6	—	0.395	199	
IU	PZT-PZNM	150 × 150 × 60	80-g steel ball dropped from 1-m height	53.1	15	
IU	PZT	304.8 × 304.8 × 21.5	Maximum applied weight of 36 kg	—	—	
IU	PZT	200 × 200 × 160	Pressing harvester	340.03	1	
IU	PZT	—	Small exciter	12.39	3.5	

DW, direct wearable; *IW*, indirect wearable; *IU*, indirect unwearable; *PZT*, piezoelectric.

2. It is best to select a suitable energy harvesting transduction mode, which has a serious impact on output performance. The rest of the mechanical functions are established based on the transduction mode.
3. Because the speeds of human gait are different during walking, it is best to design and optimize the energy harvester by considering real human walking motion rather than testing the devices at a constant excitation in experiments.
4. Apart from satisfying the requirement of technical functionality, practical feasibility and applicability should be considered according to real situations. For example, large power output is easily achieved for large-scale devices, but it is not appropriate for applications with a compact size requirement.
5. Wearable energy harvesters also need to maintain user comfort according to every aspect of mechanical design.
6. The service life of the devices is an important aspect of the design. It would be meaningless if devices lasted for only a short period, especially to supply power to equipment with low-power consumption.

2.8 Conclusion

This chapter provides an overview of energy harvesters based on their use as wearables, to exploit their characteristics and functionality from various excitation sources of human walking. Direct wearable generators produce less energy output than the other generators owing to the single vibration source. Although shoe-mounted devices are restricted by their limited size, it is possible to produce enough electricity from all excitation sources of human walking. Indirect unwearable generators with a minimal effect on user comfort demonstrate the best output performance owing to a large external pressure such as human weight, whereas extra protection mechanisms are essential in case the applied force exceeds the design limit. The main challenges of mechanical design are a large footprint size, low output performance, and low structural safety. To address these challenges, several potential applicable energy harvesting techniques are reviewed. It is best to integrate the advantages of various transduction techniques to develop a multifunctional energy harvester that can capture energy from various directions under various excitations. Moreover, it is desirable to enhance the applicability and practical feasibility of energy harvesters with a low cost and high structural safety by employing appropriate improvements with the preconditions of improving structural issues.

References

Alsharif, M. H., Kim, S., & Kuruoğlu, N. (2019). Energy harvesting techniques for wireless sensor networks/radio-frequency identification: A review. *Symmetry, 11*(7). https://doi.org/10.3390/sym11070865

Berdy, D. F., Valentino, D. J., & Peroulis, D. (2015). Kinetic energy harvesting from human walking and running using a magnetic levitation energy harvester. *Sensors and Actuators, A: Physical, 222*, 262–271. https://doi.org/10.1016/j.sna.2014.12.006

von Buren, T., Mitcheson, P. D., Green, T. C., Yeatman, E. M., Holmes, A. S., & Troster, G. (2006). Optimization of inertial micropower generators for human walking motion. *IEEE Sensors Journal, 6*(1), 28–38. https://doi.org/10.1109/JSEN.2005.853595

Cao, J., Wang, W., Zhou, S., Inman, D. J., & Lin, J. (2015). Nonlinear time-varying potential bistable energy harvesting from human motion. *Applied Physics Letters, 107*(14). https://doi.org/10.1063/1.4932947

Cao, J., Zhou, S., Wang, W., & Lin, J. (2015). Influence of potential well depth on nonlinear tristable energy harvesting. *Applied Physics Letters, 106*(17). https://doi.org/10.1063/1.4919532

Chen, C., Chau, L. Y., & Liao, W.-H. (2017). A knee-mounted biomechanical energy harvester with enhanced efficiency and safety. *Smart Materials and Structures, 26*(6).

Chen, S., Wang, L., Zhou, W., Musgrave, P., Xu, T. B., & Zuo, L. (2015). Optimal design of force magnification frame of a piezoelectric stack energy harvester. In *Proceedings of SPIE - the international society for optical engineering* (Vol. 9435)SPIE. https://doi.org/10.1117/12.2084284

Choo, J. F., Ha, D. H., & Han, H. K. (2017). Evaluation of energy-producing capability from pedestrian-induced vibration in footbridge by a new tuned mass Piezo-Damper. *KSCE Journal of Civil Engineering, 21*(6), 2322–2328. https://doi.org/10.1007/s12205-016-2817-y

Cho, H. S., Yang, J. H., Park, S. H., Yun, K. S., Kim, Y. J., & Lee, J. H. (2016). An exploratory study on the feasibility of a foot gear type energy harvester using a textile coil inductor. *Journal of Electrical Engineering and Technology, 11*(5), 1210–1215. https://doi.org/10.5370/JEET.2016.11.5.1210

Chua, Y. L., Tay, T. T., & Lim, J. Y. (2016). Electrical power harvesting using piezoelectric and human kinetic motion. *Journal of Energy and Environment, 8*(1), 11–14.

Dai, X. (2016). An vibration energy harvester with broadband and frequency-doubling characteristics based on rotary pendulums. *Sensors and Actuators, A: Physical, 241*, 161–168. https://doi.org/10.1016/j.sna.2016.02.004

Donelan, J. M., Li, Q., Naing, V., Hoffer, J. A., Weber, D. J., & Kuo, A. D. (2008). Biomechanical energy harvesting: Generating electricity during walking with minimal user effort. *Science, 319*(5864), 807–810. https://doi.org/10.1126/science.1149860

Elvin, N. G., & Elvin, A. A. (2013). Vibrational energy harvesting from human gait. *IEEE/ASME Transactions on Mechatronics, 18*(2), 637–644. https://doi.org/10.1109/TMECH.2011.2181954

Fan, K., Liu, Z., Liu, H., Wang, L., Zhu, Y., & Yu, B. (2017). Scavenging energy from human walking through a shoe-mounted piezoelectric harvester. *Applied Physics Letters, 110*(14). https://doi.org/10.1063/1.4979832

Fan, K., Yu, B., & Tang, L. (2017). Scavenging energy from human limb motions. In *Proceedings of SPIE - the international society for optical engineering* (Vol. 10164)SPIE. https://doi.org/10.1117/12.2260576

Feenstra, J., Granstrom, J., & Sodano, H. (2008). Energy harvesting through a backpack employing a mechanically amplified piezoelectric stack. *Mechanical Systems and Signal Processing, 22*(3), 721–734. https://doi.org/10.1016/j.ymssp.2007.09.015

Fu, H., Cao, K., Xu, R., Bhouri, M. A., Martinez-Botas, R., Kim, S. G., & Yeatman, E. M. (2016). Footstep energy harvesting using heel strike-induced airflow for human activity sensing. In *BSN 2016 - 13th annual body sensor networks conference* (pp. 124–129). Institute of Electrical and Electronics Engineers Inc. https://doi.org/10.1109/BSN.2016.7516245

Fu, H., Xu, R., Seto, K., Yeatman, E. M., & Kim, S. G. (2015). Energy harvesting from human motion using footstep-induced airflow. *Journal of Physics: Conference Series, 660*(1). https://doi.org/10.1088/1742-6596/660/1/012060

Fu, H., & Yeatman, E. M. (2017). A methodology for low-speed broadband rotational energy harvesting using piezoelectric transduction and frequency up-conversion. *Energy, 125*, 152–161. https://doi.org/10.1016/j.energy.2017.02.115

Goudar, V., Ren, Z., Brochu, P., Pei, Q., & Potkonjak, M. (2013). Optimizing the configuration and control of a novel human-powered energy harvesting system. In *2013 23rd international workshop on power and timing modeling, optimization and simulation, PATMOS* (pp. 75–82). IEEE Computer Society. https://doi.org/10.1109/PATMOS.2013.6662158

Granstrom, J., Feenstra, J., Sodano, H. A., & Farinholt, K. (2007). Energy harvesting from a backpack instrumented with piezoelectric shoulder straps. *Smart Materials and Structures, 16*(5), 1810–1820. https://doi.org/10.1088/0964-1726/16/5/036

Halim, M. A., & Park, J. Y. (2014a). A non-resonant, frequency up-converted electromagnetic energy harvester from human-body-induced vibration for hand-held smart system applications. *Journal of Applied Physics, 115*(9). https://doi.org/10.1063/1.4867216

Halim, M. A., & Park, J. Y. (2014b). Theoretical modeling and analysis of mechanical impact driven and frequency up-converted piezoelectric energy harvester for low-frequency and wide-bandwidth operation. *Sensors and Actuators, A: Physical, 208*, 56–65. https://doi.org/10.1016/j.sna.2013.12.033

Hamid, R., & Yuce, M. R. (2017). A wearable energy harvester unit using piezoelectric–electromagnetic hybrid technique. *Sensors and Actuators, A: Physical, 257*, 198–207. https://doi.org/10.1016/j.sna.2017.02.026

Harne, R. L., & Wang, K. W. (2013). *A review of the recent research on vibration energy harvesting via bistable systems* (Vol. 22).

He, C., Zhu, W., Chen, B., Xu, L., Jiang, T., Han, C. B., Gu, G. Q., Li, D., & Wang, Z. L. (2017). Smart floor with integrated triboelectric nanogenerator as energy harvester and motion sensor. *ACS Applied Materials and Interfaces, 9*(31), 26126–26133. https://doi.org/10.1021/acsami.7b08526

Howells, C. A. (2009). Piezoelectric energy harvesting. *Energy Conversion and Management, 50*(7), 1847–1850. https://doi.org/10.1016/j.enconman.2009.02.020

Huang, S. C., & Tsai, C. Y. (2016). Theoretical analysis of a new adjustable broadband PZT beam vibration energy harvester. *International Journal of Mechanical Sciences, 105*, 304–314. https://doi.org/10.1016/j.ijmecsci.2015.11.027

Hu, G., Tang, L., Das, R., & Marzocca, P. (2018). A two-degree-of-freedom piezoelectric energy harvester with stoppers for achieving enhanced performance. *International Journal of Mechanical Sciences, 149*, 500–507. https://doi.org/10.1016/j.ijmecsci.2017.07.051

Hwang, S. J., Jung, H. J., Kim, J. H., Ahn, J. H., Song, D., Song, Y., Lee, H. L., Moon, S. P., Park, H., & Sung, T. H. (2015). Designing and manufacturing a piezoelectric tile for harvesting energy from footsteps. *Current Applied Physics, 15*(6), 669–674. https://doi.org/10.1016/j.cap.2015.02.009

Ishida, K., Huang, T. C., Honda, K., Shinozuka, Y., Fuketa, H., Yokota, T., Zschieschang, U., Klauk, H., Tortissier, G., Sekitani, T., Toshiyoshi, H., Takamiya, M., Someya, T., & Sakurai, T. (2013). Insole pedometer with piezoelectric energy harvester and 2 v organic circuits. *IEEE Journal of Solid-State Circuits, 48*(1), 255–264. https://doi.org/10.1109/JSSC.2012.2221253

Partridge, J. S., Bucknall, R. W. G., & Kun, C. (2016). An analysis of the energy flow and energy potential from human energy harvesting with a focus on walking. *Cogent Engineering, 3*(1), 1215203. https://doi.org/10.1080/23311916.2016.1215203

Kamruzzaman, S. M., Fernando, X., & Jaseemuddin, M. (2017). Energy harvesting wireless sensors for smart cities. In *IHTC 2017 - IEEE Canada international humanitarian technology conference 2017* (pp. 218–221). Institute of Electrical and Electronics Engineers Inc. https://doi.org/10.1109/IHTC.2017.8058192

Kan, J., Fu, J., Wang, S., Zhang, Z., Chen, S., & Yang, C. (2017). Study on a piezo-disk energy harvester excited by rotary magnets. *Energy, 122*, 62–69. https://doi.org/10.1016/j.energy.2017.01.059

Kim, J. H., Hwang, S. J., Song, Y., Yang, C. H., Woo, M. S., Song, K. J., & Sung, T. H. (2016). Designing a piezoelectric energy harvester using clicking mechanism. In *2016 IEEE international conference on renewable energy research and applications, ICRERA 2016* (pp. 496–499). Institute of Electrical and Electronics Engineers Inc. https://doi.org/10.1109/ICRERA.2016.7884386

Kim, P., Nguyen, M. S., Kwon, O., Kim, Y. J., & Yoon, Y. J. (2016). Phase-dependent dynamic potential of magnetically coupled two-degree-of-freedom bistable energy harvester. *Scientific Reports, 6*. https://doi.org/10.1038/srep34411

Kim, P., Son, D., & Seok, J. (2016). Triple-well potential with a uniform depth: Advantageous aspects in designing a multi-stable energy harvester. *Applied Physics Letters, 108*(24). https://doi.org/10.1063/1.4954169

Klein, J., & Zuo, L. (2017). A velocity-amplified electromagnetic energy harvester for small amplitude vibration. *Smart Materials and Structures, 26*(9).

Kornbluh, R. D., Pelrine, R., Prahlad, H., Wong-Foy, A., McCoy, B., Kim, S., Eckerle, J., & Low, T. (2012). From boots to buoys: Promises and challenges of dielectric elastomer energy harvesting. In *Electroactivity in polymeric materials* (pp. 67–93). Springer US. https://doi.org/10.1007/978-1-4614-0878-9_3

Kuang, Y., Yang, Z., & Zhu, M. (2016). Design and characterisation of a piezoelectric knee-joint energy harvester with frequency up-conversion through magnetic plucking. *Smart Materials and Structures, 25*(8). https://doi.org/10.1088/0964-1726/25/8/085029

Kuang, Y., & Zhu, M. (2016). Characterisation of a knee-joint energy harvester powering a wireless communication sensing node. *Smart Materials and Structures, 25*(5). https://doi.org/10.1088/0964-1726/25/5/055013

Kuo, A. D. (2005). Harvesting energy by improving the economy of human walking. *Science, 309*(5741), 1686–1687. https://doi.org/10.1126/science.1118058

Kwon, D. S., Ko, H. J., & Kim, J. (2017). Piezoelectric and electromagnetic hybrid energy harvester using two cantilevers for frequency up-conversion. In *Proceedings of the IEEE international conference on micro electro mechanical systems (MEMS)* (pp. 49–52). Institute of Electrical and Electronics Engineers Inc. https://doi.org/10.1109/MEMSYS.2017.7863336

Kymissis, J., Kendall, C., Paradiso, J., & Gershenfeld, N. (1998). Parasitic power harvesting in shoes. In *Proc. of second international symposium on wearable computers* (pp. 132–139).

Lirong, W., Shubin, C., Wanlu, Z., Tian-Bing, X., & Lei, Z. (2016). Piezoelectric vibration energy harvester with two-stage force amplification. *Journal of Intelligent Material Systems and Structures*, 1175–1187. https://doi.org/10.1177/1045389X16667551

Li, H., Tian, C., & Deng, Z. D. (2014). Energy harvesting from low frequency applications using piezoelectric materials. *Applied Physics Reviews, 1*(4). https://doi.org/10.1063/1.4900845

Martin, J. P., & Li, Q. (2017). Overground vs. treadmill walking on biomechanical energy harvesting: An energetics and EMG study. *Gait and Posture, 52*, 124–128. https://doi.org/10.1016/j.gaitpost.2016.11.027

Michele, P., & Meiling, Z. (2011). Plucked piezoelectric bimorphs for knee-joint energy harvesting: Modelling and experimental validation. *Smart Materials and Structures.* https://doi.org/10.1088/0964-1726/20/5/055007

Mitcheson, P. D., Reilly, E. K., Toh, T., Wright, P. K., & Yeatman, E. M. (2007). Performance limits of the three MEMS inertial energy generator transduction types. *Journal of Micromechanics and Microengineering, 17*(9), S211–S216. https://doi.org/10.1088/0960-1317/17/9/S01

Moro, L., & Benasciutti, D. (2010). Harvested power and sensitivity analysis of vibrating shoe-mounted piezoelectric cantilevers. *Smart Materials and Structures, 19*(11). https://doi.org/10.1088/0964-1726/19/11/115011

Nathan, S., Dušan, V., & Shashank, P. (2016). Floor tile energy harvester for self-powered wireless occupancy sensing. In *Energy harvesting and systems.* https://doi.org/10.1515/ehs-2014-0009

Paul, P. J., Tutu, R. S. D., Richards, W. K., & Jerome, V. M. (2015). Project power shoe: Piezoelectric wireless power transfer - a mobile charging technique. In *Proceedings of the 5th IEEE global humanitarian technology conference, GHTC* (pp. 334–339). Institute of Electrical and Electronics Engineers Inc. https://doi.org/10.1109/GHTC.2015.7343993

Purviance, T., Wickler, S., Clayson, K., Barnes, T., & Mo, C. (2013). Development of low-profile piezoelectric energy harvester for high load application. In *2013 1st IEEE conference on technologies for sustainability, SusTech* (pp. 120–125). https://doi.org/10.1109/SusTech.2013.6617308

Purwadi, A. M., Parasuraman, S., Khan, M. K. A. A., & Elamvazuthi, I. (2015). Development of biomechanical energy harvesting device using heel strike. *Procedia Computer Science, 76*, 270–275. https://doi.org/10.1016/j.procs.2015.12.288

Rahman, M. M., Atkin, R., & Kim, H. (2015). Optimization of a microfluidic based electromagnetic energy harvester for shoe insoles. *Journal of Physics: Conference Series, 660*(1). https://doi.org/10.1088/1742-6596/660/1/012061

Rao, Y., Cheng, S., & Arnold, D. P. (2013). An energy harvesting system for passively generating power from human activities. *Journal of Micromechanics and Microengineering, 23*(11). https://doi.org/10.1088/0960-1317/23/11/114012

Rome, L. C., Flynn, L., Goldman, E. M., & Yoo, T. D. (2005). Generating electricity while walking with loads. *Science, 309*, 1725–1728. https://doi.org/10.1126/science.1111063

Saha, P., Goswami, S., Chakrabarty, S., & Sarkar, S. (2014). Simulation and model verification of shoe embedded piezoelectric energy harvester. In *Proceedings of 6th IEEE power India international conference, PIICON.* Institute of Electrical and Electronics Engineers Inc. https://doi.org/10.1109/34084POWERI.2014.7117702

Sekhar, A. C., Kishore, B. M., & Raju, T. J. (2014). Electromagnetic foot step power generation. *International Journal of Scientific and Research Publications, 4*(6), 1–5.

Semsudin, N. A. A., Sampe, J., Islam, M. S., & Md Zain, A. R. (2015). Architecture of ultra-low-power micro energy harvester using hybrid input for biomedical devices. *Asian Journal of Scientific Research, 8*(2), 212–224. https://doi.org/10.3923/ajsr.2015.212.224

Shaikh, F. K., & Zeadally, S. (2016). Energy harvesting in wireless sensor networks: A comprehensive review. *Renewable and Sustainable Energy Reviews, 55*, 1041–1054. https://doi.org/10.1016/j.rser.2015.11.010

Siddique, A. R. M., Mahmud, S., & Heyst, B. V. (2015). A comprehensive review on vibration based micro power generators using electromagnetic and piezoelectric transducer mechanisms. *Energy Conversion and Management, 106*, 728–747. https://doi.org/10.1016/j.enconman.2015.09.071

Siddique, A. R. M., Mahmud, S., & Van Heyst, B. (2017). Energy conversion by 'T-shaped' cantilever type electromagnetic vibration based micro power generator from low frequency vibration sources. *Energy Conversion and Management, 133*, 399–410. https://doi.org/10.1016/j.enconman.2016.10.059

Starner, T. (1996). Human-powered wearable computing. *IBM Systems Journal, 35*(3–4), 618–629. https://doi.org/10.1147/sj.353.0618

Starner, T., & Paradiso, J. A. (2004). Human generated power for mobile electronics. In C. Piguet (Ed.), *Low power electronics design* (pp. 1–35). CRC Press.

Valenta, C. R., & Durgin, G. D. (2014). Harvesting wireless power: Survey of energy-harvester conversion efficiency in far-field, wireless power transfer systems. *IEEE Microwave Magazine, 15*(4), 108–120. https://doi.org/10.1109/MMM.2014.2309499

Wang, W., Cao, J. Y., Cai, Y. L., Zhou, S. X., & Lin, J. (2015). Efficient energy harvesting from human motion by tristable piezoelectric cantilever. In *Proceedings of the 2015 symposium on piezoelectricity, acoustic waves and device applications, SPAWDA* (pp. 14–17). Institute of Electrical and Electronics Engineers Inc. https://doi.org/10.1109/SPAWDA.2015.7364430

Wang, W., Cao, J., Zhang, N., Lin, J., & Liao, W. H. (2017). Magnetic-spring based energy harvesting from human motions: Design, modeling and experiments. *Energy Conversion and Management, 132*, 189–197. https://doi.org/10.1016/j.enconman.2016.11.026

Wang, X., Shi, Z., Wang, J., & Xiang, H. (2016). A stack-based flex-compressive piezoelectric energy harvesting cell for large quasi-static loads. *Smart Materials and Structures, 25*(5). https://doi.org/10.1088/0964-1726/25/5/055005

Wen, S., & Xu, Q. (2018). Design of a two-stage force amplification frame for piezoelectric energy harvesting. In *2017 IEEE international conference on cybernetics and intelligent systems, CIS 2017 and IEEE conference on robotics, automation and mechatronics, RAM 2017 - proceedings* (pp. 490–495). Institute of Electrical and Electronics Engineers Inc. https://doi.org/10.1109/ICCIS.2017.8274825. Vol. 2018.

Wen, S., Xu, Q., & Zi, B. (2018). Design of a new piezoelectric energy harvester based on compound two-stage force amplification frame. *IEEE Sensors Journal, 18*(10), 3989–4000. https://doi.org/10.1109/JSEN.2018.2820221

Wu, S., Luk, P. C. K., Li, C., Zhao, X., Jiao, Z., & Shang, Y. (2017). An electromagnetic wearable 3-DoF resonance human body motion energy harvester using ferrofluid as a lubricant. *Applied Energy, 197*, 364–374. https://doi.org/10.1016/j.apenergy.2017.04.006

Xie, L., & Cai, M. (2014). Increased piezoelectric energy harvesting from human footstep motion by using an amplification mechanism. *Applied Physics Letters, 105*(14). https://doi.org/10.1063/1.4897624

Xie, L., & Cai, M. (2015a). An in-shoe harvester with motion magnification for scavenging energy from human foot strike. *IEEE/ASME Transactions on Mechatronics, 20*(6), 3264–3268. https://doi.org/10.1109/TMECH.2015.2428618

Xie, L., & Cai, M. (2015b). Development of a suspended backpack for harvesting biomechanical energy. *Journal of Mechanical Design, Transactions of the ASME, 137*(5). https://doi.org/10.1115/1.4029807

Xie, L., Li, J., Cai, S., & Li, X. (2016). Design and experiments of a self-charged power bank by harvesting sustainable human motion. *Advances in Mechanical Engineering, 8*(5), 1–10. https://doi.org/10.1177/1687814016651371

Xu, T. B., Siochi, E. J., Kang, J. H., Zuo, L., Zhou, W., Tang, X., & Jiang, X. (2013). Energy harvesting using a PZT ceramic multilayer stack. *Smart Materials and Structures, 22*(6). https://doi.org/10.1088/0964-1726/22/6/065015

Yang, W., Chen, J., Zhu, G., Yang, J., Bai, P., Su, Y., Jing, Q., Cao, X., & Wang, Z. L. (2013). Harvesting energy from the natural vibration of human walking. *ACS Nano, 7*(12), 11317–11324. https://doi.org/10.1021/nn405175z

Yang, W., & Towfighian, S. (2017). Internal resonance and low frequency vibration energy harvesting. *Smart Materials and Structures, 26*(9). https://doi.org/10.1088/1361-665X/aa791d

Ylli, K., Hoffmann, D., Willmann, A., Becker, P., Folkmer, B., & Manoli, Y. (2015). Energy harvesting from human motion: Exploiting swing and shock excitations. *Smart Materials and Structures, 24*(2). https://doi.org/10.1088/0964-1726/24/2/025029

Yulong, L., Weinong, F., Wei, L., & Mingui, S. (2016). Design and analysis of a shoe-embeded power harvester based on magnetic gear. *IEEE Transactions on Magnetics*, 1–4. https://doi.org/10.1109/TMAG.2016.2521765

Zeng, P., & Khaligh, A. (2013). A permanent-magnet linear motion driven kinetic energy harvester. *IEEE Transactions on Industrial Electronics, 60*(12), 5737–5746. https://doi.org/10.1109/TIE.2012.2229674

Zhao, S., & Erturk, A. (2013). On the stochastic excitation of monostable and bistable electroelastic power generators: Relative advantages and tradeoffs in a physical system. *Applied Physics Letters, 102*(10). https://doi.org/10.1063/1.4795296

Zhao, S., & Erturk, A. (2014). Deterministic and band-limited stochastic energy harvesting from uniaxial excitation of a multilayer piezoelectric stack. *Sensors and Actuators, A: Physical, 214*, 58–65. https://doi.org/10.1016/j.sna.2014.04.019

Zhao, N., Yang, J., Yu, Q., Zhao, J., Liu, J., Wen, Y., & Li, P. (2016). Three-dimensional piezoelectric vibration energy harvester using spiral-shaped beam with triple operating frequencies. *Review of Scientific Instruments, 87*(1). https://doi.org/10.1063/1.4940417

Zhao, J., & You, Z. (2014). A shoe-embedded piezoelectric energy harvester for wearable sensors. *Sensors (Switzerland), 14*(7), 12497–12510. https://doi.org/10.3390/s140712497

Zhou, S., Cao, J., & Lin, J. (2016). Theoretical analysis and experimental verification for improving energy harvesting performance of nonlinear monostable energy harvesters. *Nonlinear Dynamics, 86*(3), 1599–1611. https://doi.org/10.1007/s11071-016-2979-7

Zhu, G., Lin, Z. H., Jing, Q., Bai, P., Pan, C., Yang, Y., Zhou, Y., & Wang, Z. L. (2013). Toward large-scale energy harvesting by a nanoparticle-enhanced triboelectric nanogenerator. *Nano Letters, 13*(2), 847–853. https://doi.org/10.1021/nl4001053

Zi, Y., Guo, H., Wen, Z., Yeh, M. H., Hu, C., & Wang, Z. L. (2016). Harvesting low-frequency (<5 Hz) irregular mechanical energy: A possible killer application of triboelectric nanogenerator. *ACS Nano, 10*(4), 4797–4805. https://doi.org/10.1021/acsnano.6b01569

Further reading

Kim, P., & Seok, J. (2014). A multi-stable energy harvester: Dynamic modeling and bifurcation analysis. *Journal of Sound and Vibration, 333*(21), 5525–5547. https://doi.org/10.1016/j.jsv.2014.05.054

Li, X., Guo, M., & Dong, S. (2011). A flex-compressive-mode piezoelectric transducer for mechanical vibration/strain energy harvesting. *IEEE Transactions on Ultrasonics, Ferroelectrics, and Frequency Control, 58*(4), 698–703. https://doi.org/10.1109/TUFFC.2011.1862

CHAPTER 3

Survey of mechanical designs of piezoelectric energy harvester

This chapter presents a survey of mechanical designs of the piezoelectric energy harvesters. It summarizes several kinds of piezoelectric materials and reviews state-of-the-art mechanical designs of piezoelectric energy harvesters. Energy harvesters are based on different piezoelectric materials, such as piezoelectric patch, piezoelectric disc, and piezoelectric stack. The review provides a reference for designing the mechanical structure for a piezoelectric energy harvester.

3.1 Introduction

With rapid advances and developments in microelectrical-mechanical systems (MEMS) in recent decades, MEMS devices have gradually become popular in terms of sensors, actuators, controllers, and so on. Traditional energy supply devices have a number of disadvantages including a large footprint, low energy density, short working life, and high difficulty of replacement. Thus, it is difficult for them to satisfy the demands of MEMS devices. The electronics of an MEMS device have ultralow power consumption (in nano- or microwatts), which can be satisfied by a microscale energy harvester. Therefore, as a kind of self-sustaining energy supply device, microenergy harvesters have become a hot topic of research.

There are four main types of microenergy harvesters: electromagnetic (Huicong et al., 2013; Yang et al., 2009), electret (Crovetto et al., 2013; Tao et al., 2014), triboelectric (Seol et al., 2015; Zhu et al., 2013), and piezoelectric (Dhote et al., 2018; Wu et al., 2017). Piezoelectric energy harvesters have been investigated extensively because of their small footprint, simple structure, and high energy density (Liang & Liao, 2012; Shashank, 2007; Williams & Yates, 1996). Piezoelectric energy harvesters have been proposed to scavenge energy from ambient environments, including vibration (Challa et al., 2008; Fang et al., 2006; Zhao, Liu, et al., 2018), human

Mechanical Design of Piezoelectric Energy Harvesters
ISBN 978-0-12-823364-1
https://doi.org/10.1016/B978-0-12-823364-1.00006-7
© 2022 Elsevier Inc.
All rights reserved.

motion (J.G. et al., 2010; Kuang et al., 2017; Shenck & Paradiso, 2001), and flow (Crovetto et al., 2013; Tao et al., 2014).

The working principle of piezoelectric materials is based on the transformation of mechanical energy provided by input force into electrical energy, which uses the direct piezoelectric effect. The direct piezoelectric effect can be expressed as (Arnau & Soares, 2009; Moheimani & Fleming, 2006):

$$D_i = d_{ij} T_j + \varepsilon_{ik}^T E_k \tag{3.1}$$

where D_i is the dielectric displacement vector, d_{ij} is the piezoelectric stress coefficient, T_j is the stress, ε_{ik}^T is the dielectric constant, and E_k is the electric field. Subscripts i and k numbered from 1 to 3 denote Cartesian coordinate axes x, y, and z, respectively. Subscripts j numbered from 1 to 6 denote rotational motions around Cartesian coordinate axes x, y, and z, respectively. Normally, the z-axis (3-axis) is taken as the polarization direction of the piezoelectric material.

During the design of a piezoelectric energy harvester, the mechanism adopted to dispose the input force from the ambient environment is considered to be the most important part. It has a serious influence on the performance of the piezoelectric energy harvester, such as the bandwidth of resonant frequency (Stanton et al., 2010,), the safety factor (Wen et al., 2018), and the force amplification ratio (Wang, Chen, & Guzman, 2016, Wang, Shi, Wang, & Xiang, 2016).

Researchers have proposed lots of designs for piezoelectric energy harvesters. Works on piezoelectric energy harvesters has been well-reviewed in the literature (Benasciutti & Moro, 2013; Berdy et al., 2012; Caliò et al., 2014; Elahi et al., 2018; Harne & Wang, 2013; Li et al., 2014; Toprak & Tigli, 2014; Wei & Jing, 2017; Xin et al., 2016). This chapter surveys the related work and summarizes the adopted structures along with their features and functions in the designs. It provides a reference for future work on the mechanism design of a piezoelectric energy harvester.

3.2 Piezoelectric materials

In this section, three of the most common piezoelectric materials used in energy harvesters are introduced. As illustrated in Fig. 3.1, the classification is based on various shapes of piezoelectric materials, including piezoelectric patch (PEP), piezoelectric disc (PED), and piezoelectric stack (PES).

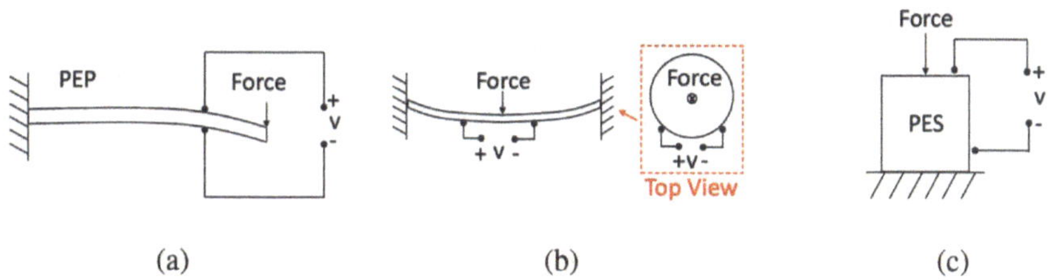

Figure 3.1 Schematic diagram of working principle for (A) piezoelectric patch (PEP), (B) piezoelectric disc (PED), and (C) piezoelectric stack (PES).

Different piezoelectric materials have own advantages and disadvantages and are most suitable for harvesting energy in different situations.

3.2.1 Piezoelectric patch

The most widely used shape of piezoelectric material in microenergy harvester is the PEP, which transforms mechanical energy into electric energy by the d31 effect. Fig. 3.1A shows a schematic diagram of the working principle for PEP. Because the stiffness of a PEP is generally low, PED-based microenergy harvesters usually have a small magnitude of resonant frequency. Therefore, as a microenergy harvester, the PEP has good performance in scavenging energy from ambient vibration with a low magnitude of input force and low input frequency. However, it is prone to break down under even a small input force, which limits its application. Moreover, the oscillation amplitude of PEPs will be rapidly reduced when the excitation frequency shifts from the resonance frequency of the mechanism. Hence, the main design objective of a PEP-based microenergy harvester is to match the resonant frequency of the mechanism with the frequency of the ambient vibration. In addition, increasing the bandwidth of the usable frequency is a good way to improve the effective frequency range of harvesting energy from ambient vibrations.

3.2.2 Piezoelectric disc

Another commonly used piezoelectric material in a microenergy harvester is PED, which transforms mechanical energy into electric energy by both d31 and d33 effects simultaneously. Fig. 3.1B shows a schematic diagram of the working principle of PEDs. Compared with PEPs, the advantage of a PED is that it can work under pressure conditions and it has a higher electromechanical transduction rate. Therefore, PEDs are usually adopted as a microenergy harvester to scavenge energy from a slowly varying input force or ambient vibration with a relatively large magnitude of input force. In addition, because the stiffness of a PED is higher than that of a PEP, a PED-based micro energy harvester may work at a higher resonance frequency.

3.2.3 Piezoelectric stack

The last commonly used piezoelectric material in a microenergy harvester is a PES, which transforms mechanical energy into electric energy by the d33 effect. Fig. 3.1C shows a schematic view, describing the working principle of PES. Compared with a PED, a PES can work under a higher-pressure condition. Because PES works based on the d33 effect, it has a much higher piezoelectric coefficient than PEP. Because the stiffness of PES is high, the main application of a PES-based micro energy harvester is scavenging energy from a larger input force. Moreover, the deformation amplitude of a PES is low owing to its high stiffness. Thus, the harvested energy is usually low if the input force is directly acting on the PES. Therefore, a force amplification mechanism is usually adopted in a PES-based microenergy harvester to enlarge the force acting on the PES. The main design objective of a PEP-based microenergy harvester is to make the force amplification ratio of the mechanism as large as possible.

3.3 Mechanism designs

For harvesting environments, there are different demands of functions for the piezoelectric energy harvester. The most common solution is to design a mechanism to realize the demanded function for the harvesting environment. Furthermore, a good mechanical design can effectively improve the performance of the piezoelectric energy harvester. In this section, common types and some special kinds of various piezoelectric materials are respectively reviewed.

3.3.1 Mechanism design of piezoelectric patch–based energy harvester

As shown in Fig. 3.2, there are five common types of mechanical design for a PEP-based microenergy harvester: normal, magnet, plectrum, stopper, and specially shaped.

The normal type is the most basic mechanical design for a PEP-based microenergy harvester. It features a cantilever beam with a proof mass. The role of the proof mass is to reduce the resonant frequency of the

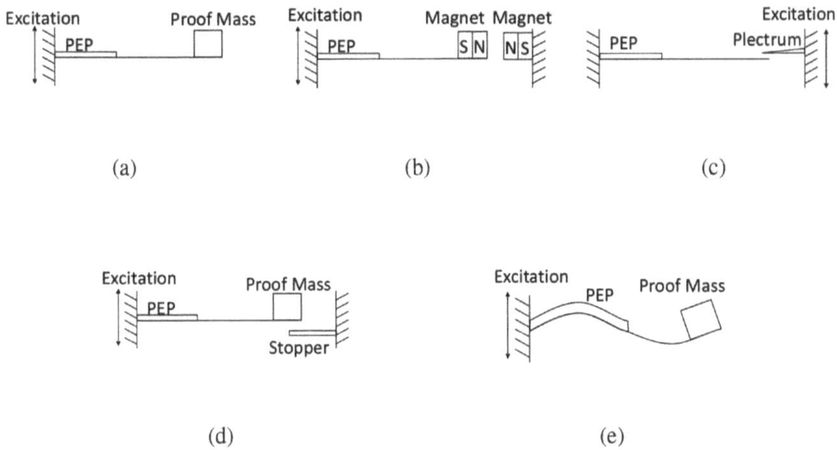

Figure 3.2 Schematic diagrams of (A) normal, (B) magnet, (C) plectrum, (D) stopper, and (E) specially shaped types of microenergy harvester based on piezoelectric patch.

mechanism to match the excitation frequency of the ambient vibration. For example, Moro et al. (Moro & Benasciutti., 2010) used a normal type shoe-mounted microenergy harvester with a PEP to scavenge energy from human walking. Chen et al. (Chen et al., 2015) designed a dandelion-like multidirectional piezoelectric vibration energy harvester (Fig. 3.3A). It was assembled using several normal types of microenergy harvesters based on PEPs with different inclination angles sensitive to excitation in different directions. Li et al. (Benasciutti & Moro, 2013; Berdy et al., 2012; Caliò et al., 2014; Elahi et al., 2018; Harne & Wang, 2013; Li et al., 2014; Toprak & Tigli, 2014; Wei & Jing, 2017; Xin et al., 2016) proposed a micro-vibration energy harvester consisting of a rectangular flexible frame and a series of normal-type mechanisms. In addition, there is a form in which the proof mass is replaced with a force receiver. It helps the cantilever beam receive the external force from an ambient environment. For instance, Kwon (Kwon, 2010), Bibo et al. (Bibo & Daqaq, 2013), and Gao et al. (Gao et al., 2013) adopted a T-shaped beam, airfoil, and cylinder, respectively, to receive external force from the flow.

A magnet type obtains nonlinear bistability by exploiting a repulsive magnetic force. Therefore, the system response will bounce between two stable states when random excitation is applied, which increases the bandwidth of useful frequency (Ferrari et al., 2010; Stanton et al., 2010). It efficiently improves the performance of a PEP-based microenergy harvester when there are wide-spectrum vibrations. Thus, there is an improved type

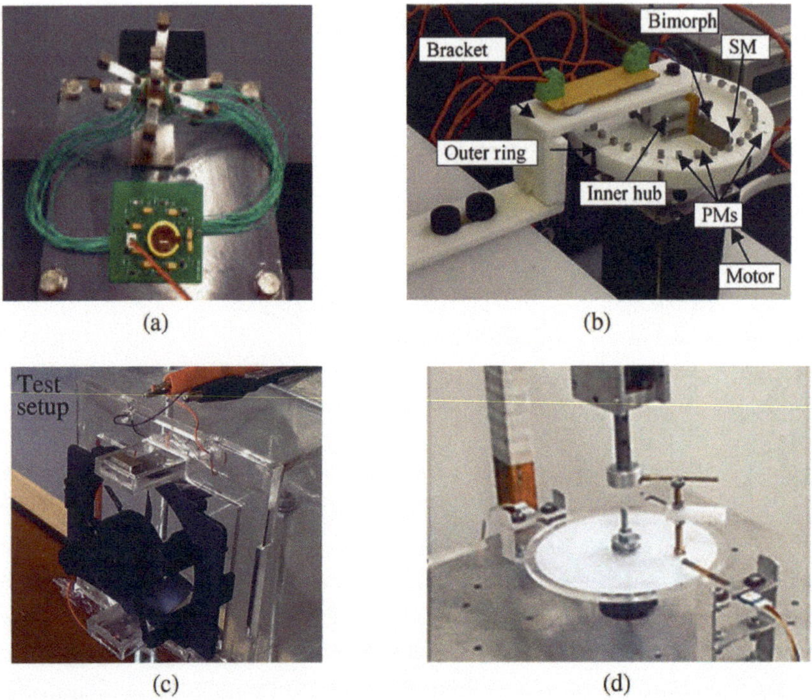

Figure 3.3 Schematic diagram of (A) a dandelion-like multidirectional piezoelectric vibration energy harvester (Chen et al., 2015); and (B–D) a magnet-type microenergy harvester to scavenge energy from rotational motion (Kuang et al., 2016; Ramezanpour et al., 2016; Rezaei-Hosseinabadi et al., 2016).

of magnet that has one magnet in the cantilever beam with multiple corresponding fixed magnets. It offers more stable states and a larger bandwidth with the useful frequency (Wang, Chen, & Guzman, 2016). Moreover, the magnet type has another use: it does not work to improve the bandwidth of the useful frequency; rather, it functions as a force transmission mechanism. The advantages of such use are free excitation end and restriction of the maximum input force for the cantilever beam, which can efficiently protect the PEP. For example, Kuang et al. (Kuang et al., 2016), Rezaei-Hosseinabadi et al. (Rezaei-Hosseinabadi et al., 2016), and Ramezanpour et al. (Ramezanpour et al., 2016) employed the magnet type of microenergy harvester to scavenge energy from rotational motion (Fig. 3.3A). Zou et al. (Gao et al., 2013; Tao et al., 2017; Zou, Zhang, Li, Gao, et al., 2017) designed a piezoelectric energy harvester consisting of two inverted cantilever beams with a magnet.

Moreover, there is a variant magnet type that adopts a ferromagnetic metal object to replace the corresponding fixed magnet. Its use is similar to that of the original magnet type. However, because magnets have the directional constraint of magnetization, the possibilities of motion and the shape of the corresponding ferromagnetic metal objects can be larger. For instance, Ferrari et al. (Ferrari et al., 2011) used a ferromagnetic cantilever instead of a magnet. Fan et al. (Fan, Chang, Chao, & Pedrycz, 2015, Fan, Chang, Pedrycz, et al., 2015; Fan, Liu, et al., 2017, Fan, Yu, et al., 2017) used a ferromagnetic ball to receive energy from various mechanical motions and the motion of human lower limbs, respectively. Fan et al. (Fan, Chang, Pedrycz, et al., 2015; Fan, Liu, et al., 2017, Fan, Yu, et al., 2017) also replaced the magnet with a ferromagnetic roller, which was used to harvest energy from sway motion.

In addition to magnets as a force transmission mechanism, the plectrum type is another such form. The working principle is that a movement plectrum will deflect the cantilever beam, and the PEP will be rapidly released. Then, it can vibrate unhindered and part of the vibration energy will be transformed into electrical energy by the piezoelectric effect. Priya (Priya, 2005), Janphuang et al. (Janphuang et al., 2015), and Kuang et al. (Kuang & Zhu, 2016) employed a PEP-based microenergy harvester and several plectra to harvest energy from rotational motion.

A stopper type of design protects the PEP from material failure (Kluger et al., 2015). It has been shown (Gu, 2011; He & Gao, 2013; Mak et al., 2011) that when a cantilever beam has a stopper, the cantilever beam will obtain a periodic impact that up-converts ambient low frequency to high frequency. Thus, power density is improved. Moreover, there is another kind of stopper type that uses a compliant cantilever beam as the stopper. It creates a coupled vibration that can significantly improve electrical power transfer and the bandwidth of the useful frequency (Challa et al., 2008; Fang et al., 2006; Zhao, Liu, et al., 2018).

The special-shaped type is a collective name for a design with a different shape of cantilever beam. Researchers have attempted to achieve better performance for microenergy harvesters by changing the shape of the cantilever beam. For example, Liu et al. (Gu & Livermore, 2011; Liu et al., 2012b; Zhao, Wang, et al., 2018) designed an S-shaped cantilever beam dedicated to ambient frequencies below 30 Hz (Fig. 3.4A). Boisseau et al. (Boisseau et al., 2013) designed a cantilever beam with an H-shaped clamped-guided beam that can increase output power in some cases (Fig. 3.4B).

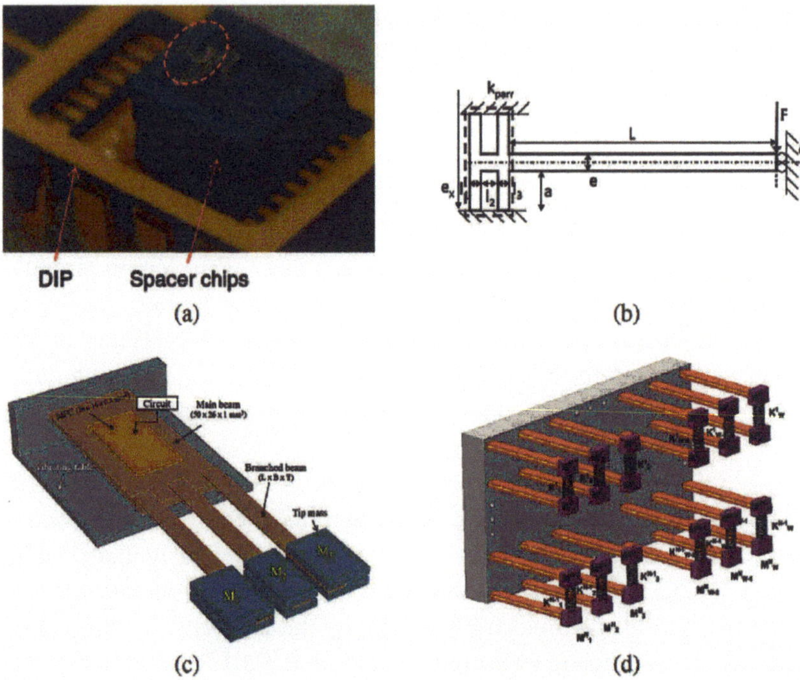

Figure 3.4 (A) An S-shaped cantilever beam (Liu et al., 2012a); (B) A cantilever beam with an H-shaped clamped-guided beam (Boisseau et al., 2013); (C) A trident-shaped energy harvester (Upadrashta & Yang, 2018); (D) A spring-parallel connected piezoelectric energy harvester (Zhang & Afzalul, 2014).

Concerning separated multiple cantilever beams, researchers have used different connection methods, including serial connection and parallel connection. Schematic diagrams for both connections are shown in Fig. 3.5. PEPs with a serial connection (H et al., 2014; Sharpes et al., 2015) or parallel

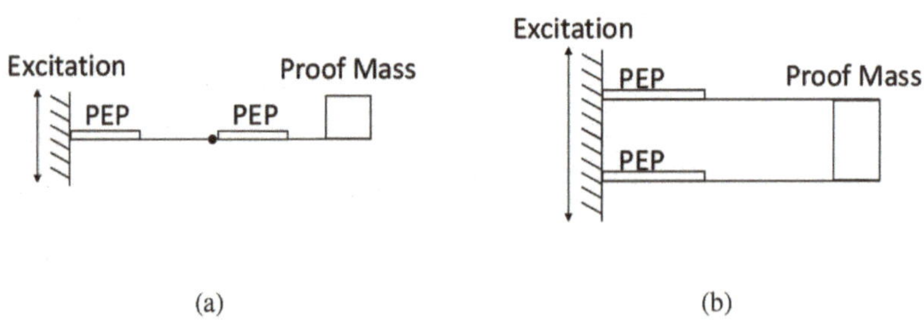

Figure 3.5 Schematic diagrams of (A) serial connection, and (B) parallel connection of microenergy harvesters with a piezoelectric patch.

connection (Hung et al., 2015; Rezaeisaray et al., 2015) of cantilevers have a wider bandwidth of useful frequency and higher output power than PEPs with a single cantilever. In addition, as shown in Fig. 3.4C, Upadrashta et al. (Upadrashta & Yang, 2018) proposed a trident-shaped piezoelectric energy harvester dedicated to broadband, low-frequency ambient vibrations. As shown in Fig. 3.4D, Zhang et al. (Zhang & Afzalul, 2014) proposed a spring-parallel connected piezoelectric energy harvester, which significantly enlarged the bandwidth of useful frequency.

Furthermore, some PEP-based microenergy harvesters possess other features. For example, Wang et al. (Wang & Tang, 2017) designed a piezoelectric energy harvester consisting of a magnet type of design serially connected to a parasitic cantilever beam. Fan et al. (Fan, Chang, Pedrycz, et al., 2015; Fan, Yu, et al., 2017) used a piezoelectric energy harvester that combined a magnet and variant magnet type. Xia et al. (Xia et al., 2017) reported a microvibration energy harvester using both piezoelectric and electromagnetic effects. Hu et al. (Hu et al., 2018) proposed a piezoelectric energy harvester of a normal type, in which the proof mass had a built-in vibratory mass and stoppers (Fig. 3.6A).

Many other PEP-based microenergy harvesters have been proposed and fabricated. Table 3.1 lists some of them and their performance.

3.3.2 Mechanism design of piezoelectric disc–based energy harvester

There are two commonly used mechanical designs for PED-based microenergy harvesters: directly acting and cymbal, as shown in Fig. 3.7.

In the directly acting type of design, excitation directly acts on the PED, which is the most primitive way. For instance, Chen et al. (Chen et al., 2015) designed a microvibration energy harvester in which a proof mass is located at the center of the PED. Ogawa et al. (Ogawa et al., 2013) and Alghisi et al. (Alghisi et al., 2015) adopted a ball to impact the PED to scavenge energy. The PED-based energy harvester proposed by Kan et al. (Kan et al., 2017) is directly driven by repulsive force between rotating magnets.

The cymbal type of design is the most common type of PED-based microenergy harvester. It is composed of one PED and two bow-shaped concave metal plates. The bow-shaped concave metal plates work as a force amplifier, which is adopted to magnify the small normal input force as a large lateral force acting on the PED. In the beginning, the cymbal type usually had a circular profile (Kim et al., 2005; Palosaari et al., 2012). Later,

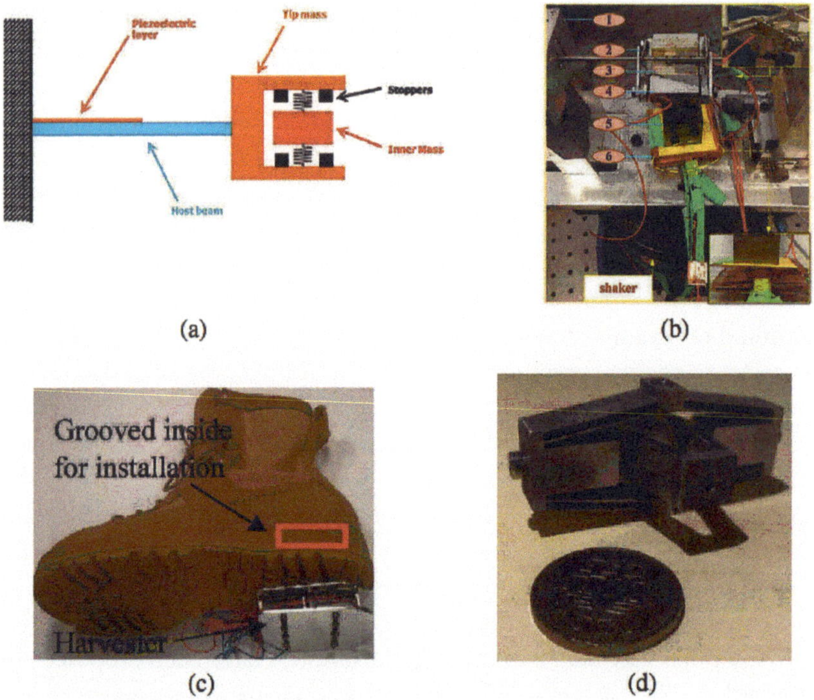

Figure 3.6 (A) A two-degree-of-freedom piezoelectric energy harvester with stoppers (Hu et al., 2018); (B) a hybrid piezoelectric-triboelectric energy harvester (Li et al., 2018); (C) a footwear energy harvester (Qian et al., 2018); (D) a two-stage rhombus force amplifier with a compound compliant hinge (Wen et al., 2018).

the rectangular profile of the cymbal type did not have periodic circumferential stress in the metal end caps, which improved the efficiency of the energy harvester; hence, it gradually became the mainstream of design (Moure et al., 2016; Nathan et al., 2016; Zou et al., 2016). In addition, Zou et al. (Zou, Zhang, Li, Gao, et al., 2017) used rotary magnets to drive the rectangular profile of the cymbal type of the PED-based microenergy harvester. Tufekcioglu and Dogan (Tufekcioglu & Dogan, 2014) designed a microvibration energy harvester in which a cantilever beam with proof mass is sandwiched between two circular profiles of cymbal types. Dedicated to ambient vibration with weak excitation and low frequency, Yang et al. (Huicong et al., 2013; Yang et al., 2009) proposed a variant of cymbal type of design, composed of one cymbal type, two elastic beams, and two

Table 3.1 Performance of piezoelectric patch–based energy harvesters.

Work	Excitation			Output		Load resistance
Chen et al. (2015)	Amplitude	0.28 mm	22 Hz	Max power	0.28 mW	8 k Ohm
Li et al. (2015)	Acceleration	1.0 g	19 Hz	Avg power	4.5 mW	600 k Ohm
Kwon (2010)	Wind	4.0 m/s	6.17 Hz	Max power	4.0 mW	4 M Ohm
Gao et al. (2015)	Wind	5.0 m/s	20.7 Hz	Avg power	30 μW	124 k Ohm
Wang et al. (2017)	Walking	7.0 km/h	–	Avg power	30.55 μW	1.3 M Ohm
Zhou et al. (2017)	Acceleration	1.0 g	5.0–16.7 Hz	Max open voltage	0.52 V	–
Kuang et al. (2016)	Typical walking	–	0.9 Hz	Avg power	5.8 mW	15 k Ohm
Zou et al. (2017)	Rotation	420 and 550 rpm	7 and 9.12 Hz	Avg power	564 and 535.3 μW	50 k Ohm
Fan et al. (2017)	Walking	8 km/h	–	RMS open voltage	1.64 V	–
Janphuang et al. (2015)	Rotation	19 rps	–	Avg power	12 μW	4.7 k Ohm
Gu (2011)	Acceleration	0.4 g	20.1 Hz	Avg power	1.53 mW	220 k Ohm
Gu & Livermore (2011)	Acceleration	0.4 g	8.2 Hz	Avg power	0.43 mW	22 k Ohm
Liu et al. (2012)	Acceleration	0.6 g	30–48 Hz	Avg power	34–100 nW	–
Zhao et al. (2018)	Acceleration	2.0 m/s^2	44.3–49 Hz	Max power	659 μW	121 k Ohm
Zhang & Afzalul (2014)	Acceleration	0.2 g	97 Hz	Avg power	136 nW	2 M Ohm
Staaf et al. (2014)	Acceleration	0.3 g	11–21 Hz	Avg power	60 μW	1 M Ohm
Fan et al. (2017)	Walking	8.0 km/h	–	Avg power	0.35 mW	1 M Ohm
Xia et al. (2017)	Acceleration	0.3 g	25.5–62 Hz	Avg power	2.78 mW	–

Avg, average; *Max*, maximum; *RMS*, Root mean square.

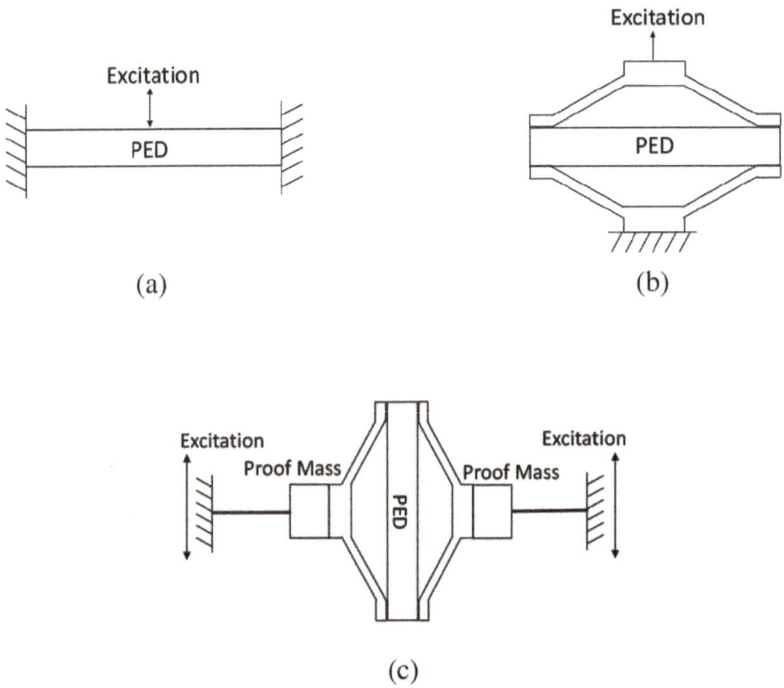

Figure 3.7 Schematic diagrams of (A) directly acting type and (B) cymbal type of microenergy harvester with piezoelectric disc, and (C) suspension cymbal type of piezoelectric energy harvester.

proof masses (see Fig. 3.7C). The elastic beams are adopted to reduce the input stiffness of the mechanism and function as the first stage of amplification. The proof masses reduce the resonant frequency of the mechanism. Moreover, Li et al. (Li et al., 2018) replaced the commonly used compliant hinge with a rigid hinge, serially connected with triboelectric patches by a spring (Fig. 3.6B).

Many other PED-based microenergy harvesters have been proposed and fabricated. Table 3.2 lists some of them and their performance.

3.3.3 Mechanism design of piezoelectric stack–based energy harvester

As shown in Fig. 3.8, there are three commonly used mechanical designs for PES-based micro energy harvesters: directly acting, rhombus force amplifier, and multistage rhombus force amplifier.

Table 3.2 Performance of piezoelectric disc–based energy harvesters.

Work	Excitation			Output		Load resistance
Alghisi et al. (2015)	Acceleration	1.0 g	113 Hz	Max power	12 mW	33 k Ohm
Kim et al. (2005)	Walking	7 km/h	–	Avg power	70 μW	1 k Ohm
Palosaari et al. (2012)	Rotation	707.5 rpm	–	Max open voltage	41.2 V	–
Sharpes et al. (2016)	Force	70 N	100 Hz	Max power	52 mW	400 k Ohm
Moure et al. (2016)	Force	24.8 N	1.19 Hz	Avg power	0.66 mW	10 M Ohm
Yang & Zu (2014)	Amplitude	12.5 mm	5 Hz	Avg power	1.5 μW	3.3 M Ohm
Zou et al. (2017)	Rotation	1000 rpm	–	Avg power	0.22 mW	240 k Ohm
Yang et al. (2015)	Acceleration	2.0 g	153 Hz	Avg power	141.6 μW	40 k Ohm
Li et al. (2017)	Acceleration	0.5 g	21 Hz	Max power	19 mW	300 k Ohm
Li et al. (2018)	Acceleration	0.5 g	26 Hz	Max power	54.7 mW	100 k Ohm
Jiang et al. (2014)	Acceleration	0.0015 g^2/Hz	10–100 Hz	RMS power	1.1 mW	300 k Ohm

Avg, average; *Max*, maximum.

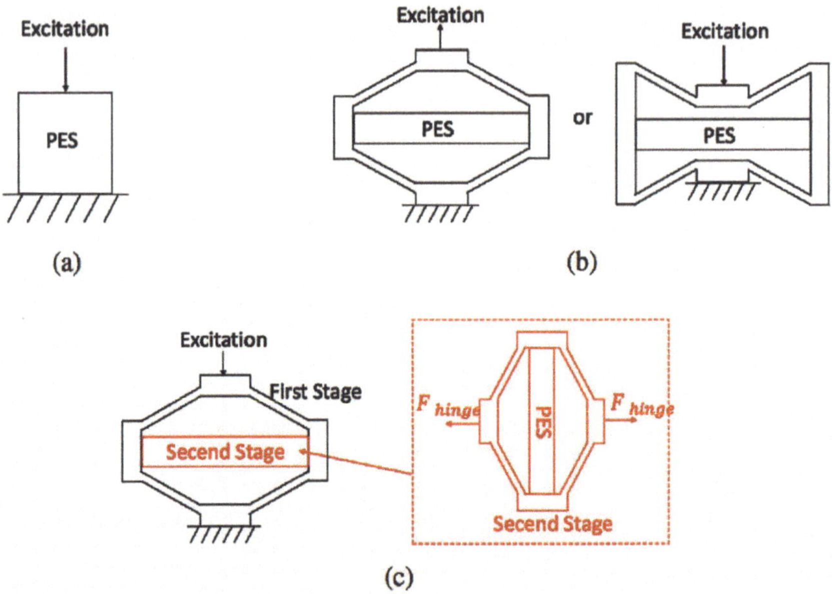

Figure 3.8 Schematic diagrams of (A) directly acting type, (B) rhombus force amplifier type, and (C) multistage rhombus force amplifier type of microenergy harvester with piezoelectric stack.

Similar to the directly acting type of PED, in the directly acting type of PES, excitation directly acts on the PES, which is the most primitive use. For instance, Jiang et al. (Xuezheng et al., 2014) and Wang et al. (Wang et al., 2015) both scavenged energy from traffic-induced pavement vibrations by PES serially connected with springs.

Because the stiffness of PES is extremely high, it is difficult to obtain deformation. Hence, it is problematic for use in energy harvesting if the acting force is not very high. To resolve this problem, a force amplifier can be introduced, which transforms a relatively low input force into a high compressive force acting on the PES. The most commonly used force amplifier in a microenergy harvester is the rhombus force amplifier. As shown in Fig. 3.8B, two kinds of rhombus force amplifiers are dedicated to pull and push input. Feenstra et al. (Feenstra et al., 2008) used a rhombus type of force amplifier to scavenge energy generated by the tensile force of a bag buckle. Li et al. (Li et al., n.d.) harvested vibration energy by a rhombus type of force amplifier with a proof mass. Bencheikh et al. (Bencheikh et al., 2014) presented a bistable broadband microenergy harvester for ambient vibration, in which a proof mass was installed between two rhombus force amplifiers and connected via compliant hinges. Wang et al. (Wang, Chen, & Guzman, 2016) designed a rhombus force amplifier type of microenergy harvester dedicated to large input loads. Chen et al. (Chen et al., 2017) built a parametric model with a rhombus type of force amplifier and found several influential parameters. Qian et al. (Qian et al., 2018) designed a footwear energy harvester connecting six rhombus types of force amplifiers in parallel (Fig. 3.6C).

The multistage rhombus type of force amplifier is an improved version of the traditional rhombus type of force amplifier. It provides a higher force amplification ratio. By inserting a rhombus force amplifier into another one, the mechanism achieves multistage force magnification. For instance, Secord et al. (Secord et al., 2010) proposed a multistage rhombus force amplifier-based energy harvester that uses both serial and parallel connections in the stages. Wang et al. (Wang, Chen, et al., 2017) presented a two-stage rhombus type of force amplifier to harvest energy from ambient vibration, in which the second stage was composed of three rhombus force amplifiers with serial connections. Wen et al. (Wen et al., 2018) designed and optimized a two-stage rhombus force amplifier with a compound compliant hinge that provided a high safety factor (Fig. 3.6D).

Many other types of force amplifiers are used in PES-based microenergy harvesters. For example, Purviance et al. (Purviance et al., 2013) designed a low-profile energy harvester with PES. The idea was inspired by the circular profile of a cymbal-type microharvester with a PED. Viet et al. (Viet et al., 2017) and Tao et al. (Tao et al., 2017) presented flow energy harvesters that used a hydraulic amplifier and lever amplifier as the force amplifier, respectively.

Lots of PES-based microenergy harvesters have been proposed and fabricated. Table 3.3 lists some of them and their performance.

Table 3.3 Performance of PES-based energy harvesters.

Work	Excitation			Output		Load resistance	Force amplification ratio
Wang et al. (2015)	Force	1360 N	6 Hz	Max power	85 mW	290 k Ohm	—
Feenstra et al. (2008)	Amplitude	4.0 mm	4 Hz	Max open voltage	25.87 V	—	—
Bencheikh et al. (2014)	Acceleration	1.0 g	87 Hz	Max power	14.6 mW	40 k Ohm	—
Wang et al. (2016)	Acceleration	0.6 g	70 Hz	Max power	2.5 mW	—	—
Chen et al. (2017)	Force	600 N	4 Hz	Max power	17.8 mW	120 k Ohm	1.72
Qian et al. (2018)	Force	100 N	1.4 Hz	Avg power	4.5 mW	30 k Ohm	8.4
Ogawa et al. (2013)	Walking	4.8 km/h	—	Avg power	20 mW	3.6 k Ohm	8.5
Wang et al. (2017)	Acceleration	—	37 Hz	Max power	2.6 W/g^2	1.7 k Ohm	21
Wen et al. (2018)	Force	2.33 N	27 Hz	Max power	0.34 mW	2 k Ohm	17.5
Purviance et al. (2013)	Force	2100 N	—	Max open voltage	34.5 V	—	—

Avg, average; *Max*, maximum.

3.4 Further discussion

From this literature review, it appears that the PEP-based microenergy harvester has been the most studied among the three commonly used piezoelectric materials owing to its flexibility. The mechanical design with the PEP tends to reduce the magnitude of resonant frequency and enlarge the bandwidth of useful frequency. Therefore, most of these are applied to scavenge energy from the low frequency of ambient vibrations. Moreover, some PEP-based microenergy harvesters are designed for other environments. There are studies on how to use the input force of environments to deform PEPs. However, because PEP material is fragile, restrictions on the maximum input force are a main design factor for the force receiver mechanism of PEPs.

Compared with PEP, the PED is more resistant and can work under pressure. Therefore, the input force can act directly on the PED. Moreover, maximization of the acting force at the PED under the broken force of a PED can efficiently improve the amount of harvested energy. Thus, the cymbal type, which can magnify the input force, is a commonly used mechanical design of PED.

The PES has the highest stiffness of the three commonly used piezoelectric materials. Therefore, the PES can directly receive an input force even under particularly high pressure. However, because of the high stiffness, it has worse performance when the acting force is not of high magnitude. Therefore, a force amplifier is widely applied to the PES-based microenergy harvester, and the force amplification ratio seriously affects the performance of the harvester.

Although the generated energy of existing microenergy harvesters with piezoelectric material is sufficient to power MEMS devices with ultralow power consumption, some devices require a higher power supply. In future work, in addition to improving existing types of mechanisms, multiple types of mechanical designs will be combined to serve as a practicable method to achieve better performance than with available designs. Furthermore, because the force amplifier will gradually become the mainstream for microenergy harvesters with PED and PES, study on improving the performance of force amplifier is important, and may be solved by designing a novel force amplifier. Moreover, because the input end of the force amplifier can receive only a fixed direction of input force, designing force transmission mechanisms dedicated to various situations is important. On the other hand, it is worth looking forward to designing a piezoelectric energy harvester mixed with other ways to generate electricity, such as solar or wind.

3.5 Conclusion

This chapter reviews state-of-the-art mechanical designs of microenergy harvesters using piezoelectric materials. Various types of piezoelectric materials are introduced and discussed and commonly used mechanical designs are presented. In addition, a discussion reveals trends in development and potential work on mechanical designs of microenergy harvesters based on piezoelectric materials. With improvements in related technologies, more and more mechanical designs for microenergy harvesters using piezoelectric materials will be proposed by both academia and industry to scavenge energy from various situations.

References

Alghisi, D., Dalola, S., Ferrari, M., & Ferrari, V. (2015). Triaxial ball-impact piezoelectric converter for autonomous sensors exploiting energy harvesting from vibrations and human motion. *Sensors and Actuators, A: Physical, 233*, 569–581. https://doi.org/10.1016/j.sna.2015.07.020

Arnau, A., & Soares, D. (2009). Fundamentals of piezoelectricity. In A. Arnau Vives (Ed.), *Piezoelectric transducers and applications* (pp. 1–38). Springer.

Benasciutti, D., & Moro, L. (2013). Energy harvesting with vibrating shoe-mounted piezoelectric cantilevers. In *Advances in energy harvesting methods* (pp. 141–162). New York: Springer. https://doi.org/10.1007/978-1-4614-5705-3_6

Bencheikh, N., Pages, A., Forissier, T., Porchez, T., Kras, A., Badel, A., & Formosa, F. (2014). *A bistable piezoelectric harvester for wideband mechanical frequency excitation.*

Berdy, D. F., Srisungsitthisunti, P., Jung, B., Xu, X., Rhoads, J. F., & Peroulis, D. (2012). Low-frequency meandering piezoelectric vibration energy harvester. *IEEE Transactions on Ultrasonics, Ferroelectrics, and Frequency Control, 59*(5), 846–858. https://doi.org/10.1109/TUFFC.2012.2269

Bibo, A., & Daqaq, M. F. (2013). Investigation of concurrent energy harvesting from ambient vibrations and wind using a single piezoelectric generator. *Applied Physics Letters, 102*(24). https://doi.org/10.1063/1.4811408

Boisseau, S., Despesse, G., & Seddik, B. A. (2013). Nonlinear h-shaped springs to improve efficiency of vibration energy harvesters. *Journal of Applied Mechanics, Transactions ASME, 80*(6). https://doi.org/10.1115/1.4023961

Caliò, R., Rongala, U. B., Camboni, D., Milazzo, M., Stefanini, C., de Petris, G., & Oddo, C. M. (2014). Piezoelectric energy harvesting solutions. *Sensors (Switzerland), 14*(3), 4755–4790. https://doi.org/10.3390/s140304755

Challa, V. R., Prasad, M., Shi, Y., & Fisher, F. T. (2008). A vibration energy harvesting device with bidirectional resonance frequency tunability. *Smart Materials and Structures, 17*(1).

Chen, R., Ren, L., Xia, H., Yuan, X., & Liu, X. (2015). Energy harvesting performance of a dandelion-like multi-directional piezoelectric vibration energy harvester. *Sensors and Actuators, A: Physical, 230*, 1–8. https://doi.org/10.1016/j.sna.2015.03.038

Chen, W., Wang, Y., & Deng, W. (2017). Deformable force amplification frame promoting piezoelectric stack energy harvesting: Parametric model, experiments and energy analysis. *Journal of Intelligent Material Systems and Structures, 28*(7), 827–836. https://doi.org/10.1177/1045389X16657426

Crovetto, A., Wang, F., & Hansen, O. (2013). An electret-based energy harvesting device with a wafer-level fabrication process. *Journal of Micromechanics and Microengineering, 23*(11). https://doi.org/10.1088/0960-1317/23/11/114010

Dhote, S., Yang, Z., & Zu, J. (2018). Modeling and experimental parametric study of a tri-leg compliant orthoplanar spring based multi-mode piezoelectric energy harvester. *Mechanical Systems and Signal Processing, 98*, 268–280. https://doi.org/10.1016/j.ymssp.2017.04.031

Elahi, H., Eugeni, M., & Gaudenzi, P. (2018). A review on mechanisms for piezoelectric-based energy harvesters. *Energies, 11*(7).

Fan, K., Chang, J., Chao, F., & Pedrycz, W. (2015). Design and development of a multi-purpose piezoelectric energy harvester. *Energy Conversion and Management, 96*, 430–439. https://doi.org/10.1016/j.enconman.2015.03.014

Fan, K., Chang, J., Pedrycz, W., Liu, Z., & Zhu, Y. (2015). A nonlinear piezoelectric energy harvester for various mechanical motions. *Applied Physics Letters, 106*(22).

Fang, H. B., Liu, J. Q., Xu, Z. Y., Dong, L., Wang, L., Chen, D., Cai, B. C., & Liu, Y. (2006). Fabrication and performance of MEMS-based piezoelectric power generator for vibration energy harvesting. *Microelectronics Journal, 37*(11), 1280–1284. https://doi.org/10.1016/j.mejo.2006.07.023

Fan, K., Liu, Z., Liu, H., Wang, L., Zhu, Y., & Yu, B. (2017). Scavenging energy from human walking through a shoe-mounted piezoelectric harvester. *Applied Physics Letters, 110*(14). https://doi.org/10.1063/1.4979832

Fan, K., Yu, B., Zhu, Y., Liu, Z., & Wang, L. (2017). Scavenging energy from the motion of human lower limbs via a piezoelectric energy harvester. *International Journal of Modern Physics B, 31*(7). https://doi.org/10.1142/S0217979217410119

Feenstra, J., Granstrom, J., & Sodano, H. (2008). Energy harvesting through a backpack employing a mechanically amplified piezoelectric stack. *Mechanical Systems and Signal Processing, 22*(3), 721–734. https://doi.org/10.1016/j.ymssp.2007.09.015

Ferrari, M., Baù, M., Guizzetti, M., & Ferrari, V. (2011). A single-magnet nonlinear piezoelectric converter for enhanced energy harvesting from random vibrations. *Sensors and Actuators, A: Physical, 172*(1), 287–292. https://doi.org/10.1016/j.sna.2011.05.019

Ferrari, M., Ferrari, V., Guizzetti, M., Andò, B., Baglio, S., & Trigona, C. (2010). Improved energy harvesting from wideband vibrations by nonlinear piezoelectric converters. *Sensors and Actuators, A: Physical, 162*(2), 425–431. https://doi.org/10.1016/j.sna.2010.05.022

Gao, X., Shih, W. H., & Shih, W. Y. (2013). Flow energy harvesting using piezoelectric cantilevers with cylindrical extension. *IEEE Transactions on Industrial Electronics, 60*(3), 1116–1118. https://doi.org/10.1109/TIE.2012.2187413

Gu, L. (2011). Low-frequency piezoelectric energy harvesting prototype suitable for the MEMS implementation. *Microelectronics Journal, 42*(2), 277–282. https://doi.org/10.1016/j.mejo.2010.10.007

Gu, L., & Livermore, C. (2011). Impact-driven, frequency up-converting coupled vibration energy harvesting device for low frequency operation. *Smart Materials and Structures, 20*(4). https://doi.org/10.1088/0964-1726/20/4/045004

Harne, R., & Wang, K. (2013). *A review of the recent research on vibration energy harvesting via bistable systems* (Vol. 22).

H, S. L. G., E, K., D, P., P, L., & P, E. (2014). Modelling and experimental verification of more efficient power harvesting by coupled piezoelectric cantilevers. *Journal of Physics: Conference Series*, 012098. https://doi.org/10.1088/1742-6596/557/1/012098

He, X. F., & Gao, J. (2013). Wind energy harvesting based on flow-induced-vibration and impact. *Microelectronic Engineering, 111*, 82–86. https://doi.org/10.1016/j.mee.2013.02.009

Huicong, L., You, Q., & Chengkuo, L. (2013). A multi-frequency vibration-based MEMS electromagnetic energy harvesting device. *Sensors and Actuators A: Physical*, 37–43. https://doi.org/10.1016/j.sna.2013.09.015

Hung, C. F., Chung, T. K., Yeh, P. C., Chen, C. C., Wang, C. M., & Lin, S. H. (2015). A miniature mechanical-piezoelectric-configured three-axis vibrational energy harvester. *IEEE Sensors Journal, 15*(10), 5601–5615. https://doi.org/10.1109/JSEN.2015.2444993

Hu, G., Tang, L., Das, R., & Marzocca, P. (2018). A two-degree-of-freedom piezoelectric energy harvester with stoppers for achieving enhanced performance. *International Journal of Mechanical Sciences, 149*, 500–507. https://doi.org/10.1016/j.ijmecsci.2017.07.051

Janphuang, P., Lockhart, R. A., Isarakorn, D., Henein, S., Briand, D., & De Rooij, N. F. (2015). Harvesting energy from a rotating gear using an AFM-Like MEMS piezoelectric frequency up-converting energy harvester. *Journal of Microelectromechanical Systems, 24*(3), 742–754. https://doi.org/10.1109/JMEMS.2014.2349794

J.G., R., L.M., G., P.F., R., M.P., S., & S., L.-M. (2010). Energy harvesting from piezoelectric materials fully integrated in footwear. *IEEE Transactions on Industrial Electronics*, 813–819. https://doi.org/10.1109/TIE.2009.2028360

Kan, J., Fu, J., Wang, S., Zhang, Z., Chen, S., & Yang, C. (2017). Study on a piezo-disk energy harvester excited by rotary magnets. *Energy, 122*, 62–69. https://doi.org/10.1016/j.energy.2017.01.059

Kim, H. W., Priya, S., Uchino, K., & Newnham, R. E. (2005). Piezoelectric energy harvesting under high pre-stressed cyclic vibrations. *Journal of Electroceramics, 15*(1), 27–34. https://doi.org/10.1007/s10832-005-0897-z

Kluger, J. M., Sapsis, T. P., & Slocum, A. H. (2015). Robust energy harvesting from walking vibrations by means of nonlinear cantilever beams. *Journal of Sound and Vibration, 341*, 174–194. https://doi.org/10.1016/j.jsv.2014.11.035

Kuang, Y., Daniels, A., & Zhu, M. (2017). A sandwiched piezoelectric transducer with flex end-caps for energy harvesting in large force environments. *Journal of Physics D: Applied Physics, 50*(34).

Kuang, Y., Yang, Z., & Zhu, M. (2016). Design and characterisation of a piezoelectric knee-joint energy harvester with frequency up-conversion through magnetic plucking. *Smart Materials and Structures, 25*(8). https://doi.org/10.1088/0964-1726/25/8/085029

Kuang, Y., & Zhu, M. (2016). Characterisation of a knee-joint energy harvester powering a wireless communication sensing node. *Smart Materials and Structures, 25*(5). https://doi.org/10.1088/0964-1726/25/5/055013

Kwon, S.-D. (2010). A t-shaped piezoelectric cantilever for fluid energy harvesting. *Applied Physics Letters, 97*(16).

Liang, J., & Liao, W. H. (2012). Impedance modeling and analysis for piezoelectric energy harvesting systems. *IEEE/ASME Transactions on Mechatronics, 17*(6), 1145–1157. https://doi.org/10.1109/TMECH.2011.2160275

Li, X., Guo, M., & Dong, S. (n.d.). A flex-compressive-mode piezoelectric transducer for mechanical vibration/strain energy harvesting. *IEEE Transactions on Ultrasonics, Ferroelectrics, and Frequency Control, 58*(4), 698–703.

Li, Z., Saadatnia, Z., Yang, Z., & Naguib, H. (2018). A hybrid piezoelectric-triboelectric generator for low-frequency and broad-bandwidth energy harvesting. *Energy Conversion and Management, 174*, 188–197. https://doi.org/10.1016/j.enconman.2018.08.018

Li, H., Tian, C., & Deng, Z. D. (2014). Energy harvesting from low frequency applications using piezoelectric materials. *Applied Physics Reviews, 1*(4). https://doi.org/10.1063/1.4900845

Liu, H., Lee, C., Kobayashi, T., Tay, C. J., & Quan, C. (2012a). A new S-shaped MEMS PZT cantilever for energy harvesting from low frequency vibrations below 30 Hz. *Microsystem Technologies, 18*(4), 497–506. https://doi.org/10.1007/s00542-012-1424-1

Liu, H., Lee, C., Kobayashi, T., Tay, C. J., & Quan, C. (2012b). Investigation of a MEMS piezoelectric energy harvester system with a frequency-widened-bandwidth mechanism introduced by mechanical stoppers. *Smart Materials and Structures, 21*(3). https://doi.org/10.1088/0964-1726/21/3/035005

Mak, K. H., McWilliam, S., Popov, A. A., & Fox, C. H. J. (2011). Performance of a cantilever piezoelectric energy harvester impacting a bump stop. *Journal of Sound and Vibration, 330*(25), 6184–6202. https://doi.org/10.1016/j.jsv.2011.07.008

Moheimani, S. R., & Fleming, A. J. (2006). *Piezoelectric transducers for vibration control and damping* (pp. 9–35).

Moro, L., & Benasciutti, D. (2010). Harvested power and sensitivity analysis of vibrating shoe-mounted piezoelectric cantilevers. *Smart Materials and Structures, 19*(11). https://doi.org/10.1088/0964-1726/19/11/115011

Moure, A., Izquierdo Rodríguez, M. A., Rueda, S. H., Gonzalo, A., Rubio-Marcos, F., Cuadros, D. U., Pérez-Lepe, A., & Fernández, J. F. (2016). Feasible integration in asphalt of piezoelectric cymbals for vibration energy harvesting. *Energy Conversion and Management, 112*, 246–253. https://doi.org/10.1016/j.enconman.2016.01.030

Nathan, S., Dušan, V., & Shashank, P. (2016). Floor tile energy harvester for self-powered wireless occupancy sensing. In *Energy harvesting and systems*. https://doi.org/10.1515/ehs-2014-0009

Ogawa, T., Sugisawa, R., Sakurada, Y., Aoshima, H., Hikida, M., & Akaishi, H. (2013). Energy harvesting devices utilizing resonance vibration of piezoelectric buzzer. *Japanese Journal of Applied Physics, 52*(9). https://doi.org/10.7567/JJAP.52.09KD14

Palosaari, J., Leinonen, M., Hannu, J., Juuti, J., & Jantunen, H. (2012). Energy harvesting with a cymbal type piezoelectric transducer from low frequency compression. *Journal of Electroceramics, 28*(4), 214–219. https://doi.org/10.1007/s10832-012-9713-8

Priya, S. (2005). Modeling of electric energy harvesting using piezoelectric windmill. *Applied Physics Letters, 87*(18), 1–3. https://doi.org/10.1063/1.2119410

Purviance, T., Wickler, S., Clayson, K., Barnes, T., & Mo, C. (2013). Development of low-profile piezoelectric energy harvester for high load application. In *2013 1st IEEE conference on technologies for sustainability, SusTech* (pp. 120–125). https://doi.org/10.1109/SusTech.2013.6617308

Qian, F., Xu, T. B., & Zuo, L. (2018). Design, optimization, modeling and testing of a piezoelectric footwear energy harvester. *Energy Conversion and Management, 171*, 1352–1364. https://doi.org/10.1016/j.enconman.2018.06.069

Ramezanpour, R., Nahvi, H., & Ziaei-Rad, S. (2016). Electromechanical behavior of a pendulum-based piezoelectric frequency up-converting energy harvester. *Journal of Sound and Vibration, 370*, 280–305. https://doi.org/10.1016/j.jsv.2016.01.052

Rezaei-Hosseinabadi, N., Tabesh, A., & Dehghani, R. (2016). A topology and design optimization method for wideband piezoelectric wind energy harvesters. *IEEE Transactions on Industrial Electronics, 63*(4), 2165–2173. https://doi.org/10.1109/TIE.2015.2499248

Rezaeisaray, M., Gowini, M. E., Sameoto, D., Raboud, D., & Moussa, W. (2015). Low frequency piezoelectric energy harvesting at multi vibration mode shapes. *Sensors and Actuators, A: Physical, 228*, 104–111. https://doi.org/10.1016/j.sna.2015.02.036

Secord, T. W., Mazumdar, A., & Asada, H. H. (2010). A multi-cell piezoelectric device for tunable resonance actuation and energy harvesting. In *Proceedings - IEEE international conference on robotics and automation* (pp. 2169–2176). https://doi.org/10.1109/ROBOT.2010.5509158

Seol, M. L., Woo, J. H., Jeon, S. B., Kim, D., Park, S. J., Hur, J., & Choi, Y. K. (2015). Vertically stacked thin triboelectric nanogenerator for wind energy harvesting. *Nano Energy, 14*, 201–208. https://doi.org/10.1016/j.nanoen.2014.11.016

Sharpes, N., Abdelkefi, A., & Priya, S. (2015). Two-dimensional concentrated-stress low-frequency piezoelectric vibration energy harvesters. *Applied Physics Letters, 107*(9). https://doi.org/10.1063/1.4929844

Shashank, P. (2007). Advances in energy harvesting using low profile piezoelectric transducers. *Journal of Electroceramics*, 167–184. https://doi.org/10.1007/s10832-007-9043-4

Shenck, N. S., & Paradiso, J. A. (2001). Energy scavenging with shoe-mounted piezoelectrics. *IEEE Micro, 21*(3), 30–42. https://doi.org/10.1109/40.928763

Stanton, S. C., McGehee, C. C., & Mann, B. P. (2010). Nonlinear dynamics for broadband energy harvesting: Investigation of a bistable piezoelectric inertial generator. *Physica D: Nonlinear Phenomena, 239*(10), 640–653. https://doi.org/10.1016/j.physd.2010.01.019

Tao, K., Liu, S., Lye, S. W., Miao, J., & Hu, X. (2014). A three-dimensional electret-based micro power generator for low-level ambient vibrational energy harvesting. *Journal of Micromechanics and Microengineering, 24*(6).

Tao, J. X., Viet, N. V., Carpinteri, A., & Wang, Q. (2017). Energy harvesting from wind by a piezoelectric harvester. *Engineering Structures, 133*, 74–80. https://doi.org/10.1016/j.engstruct.2016.12.021

Toprak, A., & Tigli, O. (2014). Piezoelectric energy harvesting: State-of-the-art and challenges. *Applied Physics Reviews, 1*(3). https://doi.org/10.1063/1.4896166

Tufekcioglu, E., & Dogan, A. (2014). A flextensional piezo-composite structure for energy harvesting applications. *Sensors and Actuators, A: Physical, 216*, 355–363. https://doi.org/10.1016/j.sna.2014.06.001

Upadrashta, D., & Yang, Y. (2018). Trident-shaped multimodal piezoelectric energy harvester. *Journal of Aerospace Engineering, 31*(5). https://doi.org/10.1061/(ASCE)AS.1943-5525.0000899

Viet, N. V., Al-Qutayri, M., Liew, K. M., & Wang, Q. (2017). An octo-generator for energy harvesting based on the piezoelectric effect. *Applied Ocean Research, 64*, 128–134. https://doi.org/10.1016/j.apor.2017.02.007

Wang, Y., Chen, W., & Guzman, P. (2016). Piezoelectric stack energy harvesting with a force amplification frame: Modeling and experiment. *Journal of Intelligent Material Systems and Structures, 27*(17), 2324–2332. https://doi.org/10.1177/1045389X16629568

Wang, L., Chen, S., Zhou, W., Xu, T. B., & Zuo, L. (2017). Piezoelectric vibration energy harvester with two-stage force amplification. *Journal of Intelligent Material Systems and Structures, 28*(9), 1175–1187. https://doi.org/10.1177/1045389X16667551

Wang, X., Shi, Z., Wang, J., & Xiang, H. (2016). A stack-based flex-compressive piezoelectric energy harvesting cell for large quasi-static loads. *Smart Materials and Structures, 25*(5).

Wang, J., Shi, Z., Xiang, H., & Song, G. (2015). Modeling on energy harvesting from a railway system using piezoelectric transducers. *Smart Materials and Structures, 24*(10). https://doi.org/10.1088/0964-1726/24/10/105017

Wang, H., & Tang, L. (2017). Modeling and experiment of bistable two-degree-of-freedom energy harvester with magnetic coupling. *Mechanical Systems and Signal Processing, 86*, 29–39. https://doi.org/10.1016/j.ymssp.2016.10.001

Wei, C., & Jing, X. (2017). A comprehensive review on vibration energy harvesting: Modelling and realization. *Renewable and Sustainable Energy Reviews, 74*, 1–18. https://doi.org/10.1016/j.rser.2017.01.073

Wen, S., Xu, Q., & Zi, B. (2018). Design of a new piezoelectric energy harvester based on compound two-stage force amplification frame. *IEEE Sensors Journal, 18*(10), 3989–4000. https://doi.org/10.1109/JSEN.2018.2820221

Williams, C. B., & Yates, R. B. (1996). Analysis of a micro-electric generator for microsystems. *Sensors and Actuators, A: Physical, 52*(1–3), 8–11. https://doi.org/10.1016/0924-4247(96)80118-X

Wu, Y., Ji, H., Qiu, J., & Han, L. (2017). A 2-degree-of-freedom cubic nonlinear piezoelectric harvester intended for practical low-frequency vibration. *Sensors and Actuators, A: Physical, 264*, 1–10. https://doi.org/10.1016/j.sna.2017.06.029

Xia, H., Chen, R., & Ren, L. (2017). Parameter tuning of piezoelectric–electromagnetic hybrid vibration energy harvester by magnetic force: Modeling and experiment. *Sensors and Actuators, A: Physical, 257*, 73–83. https://doi.org/10.1016/j.sna.2017.01.026

Xin, Y., Li, X., Tian, H., Guo, C., Qian, C., Wang, S., & Wang, C. (2016). Shoes-equipped piezoelectric transducer for energy harvesting: A brief review. *Ferroelectrics, 493*(1), 12–24. https://doi.org/10.1080/00150193.2016.1123061

Xuezheng, J., Yancheng, L., Jianchun, L., Jiong, W., & Jin, Y. (2014). Piezoelectric energy harvesting from traffic-induced pavement vibrations. *Journal of Renewable and Sustainable Energy*, 043110. https://doi.org/10.1063/1.4891169

Yang, B., Lee, C., Xiang, W., Xie, J., Han He, J., Kotlanka, R. K., Low, S. P., & Feng, H. (2009). Electromagnetic energy harvesting from vibrations of multiple frequencies. *Journal of Micromechanics and Microengineering, 19*(3). https://doi.org/10.1088/0960-1317/19/3/035001

Zhang, H., & Afzalul, K. (2014). Design and analysis of a connected broadband multi-piezoelectric-bimorph- beam energy harvester. *IEEE Transactions on Ultrasonics, Ferroelectrics, and Frequency Control, 61*(6), 1016–1023. https://doi.org/10.1109/TUFFC.2014.2997

Zhao, D., Liu, S., Xu, Q., Sun, W., Wang, T., & Cheng, Q. (2018). Theoretical modeling and analysis of a 2-degree-of-freedom hybrid piezoelectric—electromagnetic vibration energy harvester with a driven beam. *Journal of Intelligent Material Systems and Structures, 29*(11), 2465–2476. https://doi.org/10.1177/1045389X18770870

Zhao, D., Wang, X., Cheng, Y., Liu, S., Wu, Y., Chai, L., Liu, Y., & Cheng, Q. (2018). Analysis of single degree-of-freedom piezoelectric energy harvester with stopper by incremental harmonic balance method. *Materials Research Express, 5*(5).

Zhu, G., Lin, Z. H., Jing, Q., Bai, P., Pan, C., Yang, Y., Zhou, Y., & Wang, Z. L. (2013). Toward large-scale energy harvesting by a nanoparticle-enhanced triboelectric nanogenerator. *Nano Letters, 13*(2), 847–853. https://doi.org/10.1021/nl4001053

Zou, H. X., Zhang, W. M., Li, W. B., Gao, Q. H., Wei, K. X., Peng, Z. K., & Meng, G. (2017). Design, modeling and experimental investigation of a magnetically coupled flextensional rotation energy harvester. *Smart Materials and Structures, 26*(11). https://doi.org/10.1088/1361-665X/aa8eb8

Zou, H. X., Zhang, W. M., Wei, K. X., Li, W. B., Peng, Z. K., & Meng, G. (2016). Design and analysis of a piezoelectric vibration energy harvester using rolling mechanism. *Journal of Vibration and Acoustics, Transactions of the ASME, 138*(5). https://doi.org/10.1115/1.4033193

Further reading

Chen, X. R., Yang, T. Q., Wang, W., & Yao, X. (2012). Vibration energy harvesting with a clamped piezoelectric circular diaphragm. *Ceramics International, 38*(1), S271–S274. https://doi.org/10.1016/j.ceramint.2011.04.099

Haitao, L., Weiyang, Q., Chunbo, L., Wangzheng, D., & Zhiyong, Z. (2015). Dynamics and coherence resonance of tri-stable energy harvesting system. *Smart Materials and Structures, 25*(1). https://doi.org/10.1088/0964-1726/25/1/015001

Lan, C., & Qin, W. (2017). Enhancing ability of harvesting energy from random vibration by decreasing the potential barrier of bistable harvester. *Mechanical Systems and Signal Processing, 85*, 71–81. https://doi.org/10.1016/j.ymssp.2016.07.047

Li, P., Liu, Y., Wang, Y., Luo, C., Li, G., Hu, J., Liu, W., & Zhang, W. (2015). Low-frequency and wideband vibration energy harvester with flexible frame and interdigital structure. *AIP Advances, 5*(4). https://doi.org/10.1063/1.4919711

T, L. H., Z, Y., J, Z., & Y, Q. W. (2017). Numerical and experimental study of a compressive-mode energy harvester under random excitations. *Smart Materials and Structures*, 035064. https://doi.org/10.1088/1361-665X/aa5e61

Wang, W., Cao, J., Bowen, C. R., Zhou, S., & Lin, J. (2017). Optimum resistance analysis and experimental verification of nonlinear piezoelectric energy harvesting from human motions. *Energy, 118*, 221–230. https://doi.org/10.1016/j.energy.2016.12.035

Yang, Z., Zhu, Y., & Zu, J. (2015). Theoretical and experimental investigation of a nonlinear compressive-mode energy harvester with high power output under weak excitations. *Smart Materials and Structures, 24*(2). https://doi.org/10.1088/0964-1726/24/2/025028

Yang, Z., & Zu, J. (2014). High-efficiency compressive-mode energy harvester enhanced by a multi-stage force amplification mechanism. *Energy Conversion and Management, 88*, 829–833. https://doi.org/10.1016/j.enconman.2014.09.026

Zhou, Z., Qin, W., & Zhu, P. (2017). A broadband quad-stable energy harvester and its advantages over bi-stable harvester: Simulation and experiment verification. *Mechanical Systems and Signal Processing, 84*, 158–168. https://doi.org/10.1016/j.ymssp.2016.07.001

Zou, H. X., Zhang, W.m., Li, W. B., Wei, K. X., Gao, Q. H., Peng, Z. K., & Meng, G. (2017). Design and experimental investigation of a magnetically coupled vibration energy harvester using two inverted piezoelectric cantilever beams for rotational motion. *Energy Conversion and Management, 148*, 1391–1398. https://doi.org/10.1016/j.enconman.2017.07.005

CHAPTER 4

Design of a piezoelectric energy harvester based on two-stage force amplification frame

This chapter presents the design and testing of a piezoelectric energy harvester with a compound two-stage force amplification frame. It possesses a compact footprint size and achieves high power output using a single piezoelectric stack. The structural parameters are optimally designed to obtain the largest force amplification ratio to achieve the maximum power output under constraints. A simulation study is carried out by finite element analysis to demonstrate the advantage of the proposed compound two-stage force amplifier compared with a conventional design. A prototype harvester is fabricated for experimental verification. The results indicate that the proposed harvester features a high safety factor, large amplification ratio, and compact size, which can be adopted for practical application to scavenge energy induced by human footsteps during walking.

4.1 Introduction

As a monolithic ceramic structure with multiple layers of piezoelectric material, a piezoelectric stack has high compressive strength and can bear a large compressive force. To exploit the benefit of a piezoelectric stack fully, a promising solution to implement a force amplification frame to magnify force exerted on the piezoelectric stack, to harvest more power (Wang et al., 2017). The motivation for this work was to develop a force amplification mechanism to magnify the force to be high as possible with a limited physical size. Specifically, the developed harvester is dedicated to harvesting energy from human footsteps in crowded areas, such as a sports field or shopping center.

A compound force amplification frame is developed in this chapter to enhance the stability of the frame under a large loading situation. An enhanced bridge-type force amplifier is proposed by introducing multiple

Mechanical Design of Piezoelectric Energy Harvesters
ISBN 978-0-12-823364-1
https://doi.org/10.1016/B978-0-12-823364-1.00012-2
© 2022 Elsevier Inc.
All rights reserved.

beams for each of the four arms. Moreover, a compound two-stage force amplification frame with a piezoelectric stack is developed to harvest energy from human strikes. The analytical model is established based on the Euler-Bernoulli beam theory to predict the static properties of the mechanism. The design variables of the frame are determined by the analytical model, and parameters are optimized by adopting the multiobjective genetic algorithm method. Finite element analysis (FEA) simulation and experimental studies are conducted to validate the performance of the force amplifier-based piezoelectric energy harvester.

4.2 Mechanical design

This section presents the mechanical design of the piezoelectric energy harvester. It involves selecting the material, frame type, and beam type for the force amplifier mechanism.

4.2.1 Mechanism design of the frame

The proposed harvester has a compound two-stage force amplification frame; its schematic diagram is shown in Fig. 4.1. The operation principle of the frame is explained as follows. Assume that an input load force (F) is exerted vertically downward on the top of outer frame. Then, the device

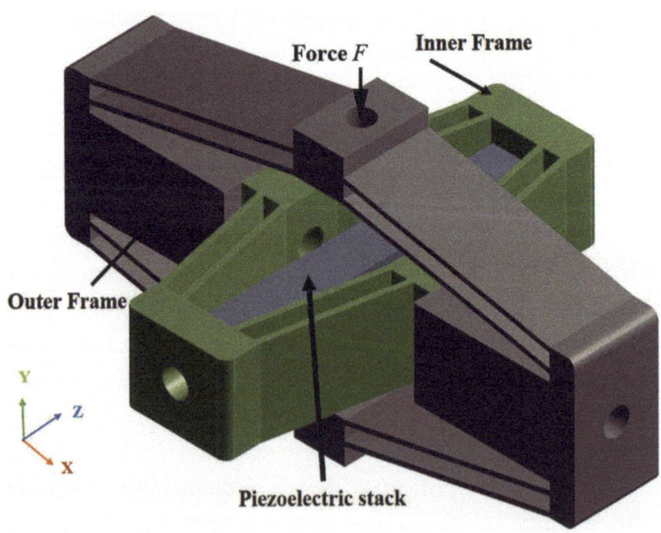

Figure 4.1 Computer-aided design (CAD) model of the proposed compound two-stage force amplification frame.

produces a horizontal tensile force along the X direction, which pulls the two output ends of the outer frame away from the center. Consider the input ends of the inner frame are connected to the output ends of the outer frame; hence, the created tensile force also pulls the two input ends of the inner frame away from the center. As a consequence, a compressive force is produced by the two output ends of the inner frame along the Z direction. This force is applied on the two ends of the embedded piezoelectric stack. Compared with the traditional bridge-type force amplification mechanism, the designed force amplifier exhibits a larger amplification ratio, higher safety factor, and more compact physical size. Such advantages are caused by the introduction of compound beams for each of the four arms, which will be verified later.

4.2.2 Selection of material

In the mechanical design stage for the force amplifier, the selection of material is crucial. In this work, the alloy aluminum 7075 is adopted. It has a higher ratio of E/ρ, in which E represents the Young's modulus and ρ indicates the material density. A higher E/ρ value enables a larger allowable deformation of the material. In addition, it is sufficient to prevent damage upon application of loads imposed by human walking. The material specifications are given in Table 4.1.

Concerning the piezoelectric stack material, a type of piezoelectric stack (Model P-885.91, Physik Instrumente Co., Ltd.) is selected. It has a compact footprint size and satisfactory d_{33} performance. The dominant physical and electromechanical properties of the piezoelectric stack material are listed in Table 4.2.

4.2.3 Selection of frame type

Typically, the force amplification frame can be categorized into two types:, convex and concave. The convex frame works with a tensile force, as demonstrated in the literature (Feenstra et al., 2008; Zhou & Zuo, 2013).

Table 4.1 Mechanical properties of aluminum 7075 alloy.

Property	Value	Unit
Density	2810	kg/m^3
Poisson's ratio	0.33	—
Young's modulus	71.7	GPa
Tensile yield strength	503	MPa

Table 4.2 Physical and electromechanical properties of piezoelectric stack.

Parameter	Symbol	Value	Unit
Relative permittivity	ε33T/ε0	2400	—
Coupling factor	k33	0.69	—
Piezoelectric charge coefficient	d33	400×10^{-12}	C/N
Piezoelectric voltage coefficient	g33	25×10^{-3}	Vm/N
Elastic compliance coefficient	S33E	20.7×10^{-12}	m^2/N
Capacitance	C	3.1×10^{-6}	F

By contrast, the concave frame operates with a compressive force (Chen et al., 2017; Wang et al., 2016). Each type of frame has specific advantages.

Considering a single amplification frame, the convex frame generally provides a larger force amplification ratio than the concave one because of a large deformation in the input end. The large deformation results in a small safety factor because the bending area is increased. In addition, regarding the mechanism design of the two-stage force amplification, the input force is amplified in the operation process, whereas the output displacement of the frame is reduced in view of the law of conservation of energy.

As for the concave frame, the magnitude of the reduced output displacement of the outer frame is insufficient for the inner frame to produce a large amplification ratio. Most of strain energy is stored in the frame to produce mechanical deformation. As a result, the total amplification ratio of the entire frame becomes small. On the contrary, the convex frame has a low deformation of the input end, which improves the energy conversion efficiency and safety. Strain energy stored in the outer and inner frame beams produces mechanical deformation. Usually, a large deformation arises from large bending of the elastic beams, which induces an influence on energy conversion efficiency and safety of the mechanism.

Considering the two types of force amplification frames with identical parameters, the simulation results of stress distribution are shown in Fig. 4.2. The output performances are compared, as given in Table 4.3. Because the convex frame has a larger force amplification ratio with a higher safety factor, the convex frame was adopted for further investigations.

Compared with a two-dimensional (2D) simulation study, the settings in a 3D simulation study are closer to conditions in a real situation. Therefore, all mechanical design optimizations were carried out based on the 3D simulation to ensure the consistency of results in this work.

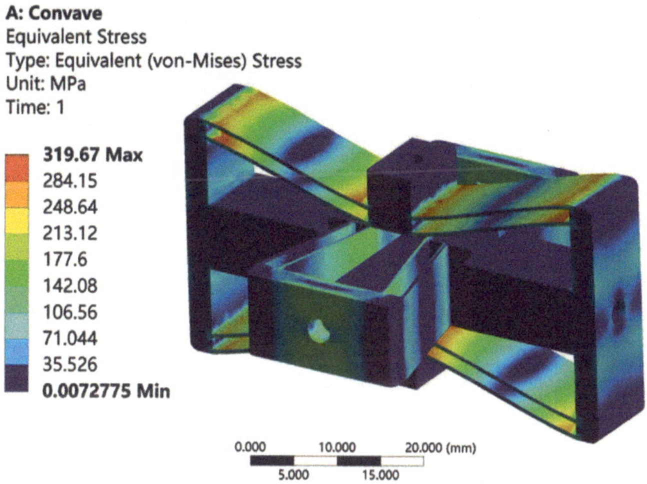

(a)

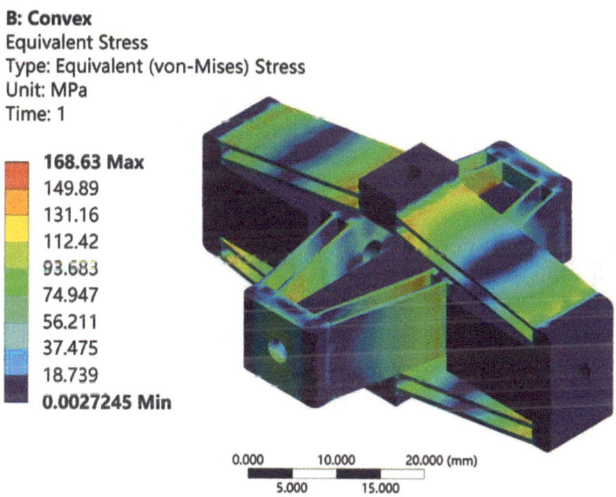

(b)

Figure 4.2 Stress distribution based on safety for two types of two-stage amplification frames: (A) concave type; (B) convex type.

4.2.4 Selection of beam type

The bridge-type amplification mechanism has been reported using various structures. For example, the conventional beam-based bridge-type frame (with beams only) was investigated (Chen et al., 2017; Feenstra et al., 2008;

Table 4.3 Performance of concave frame and convex frame under input force of 100 N.

Performance	Two-stage concave frame	Two-stage convex frame
Beam deflection (mm)	3.07	1.05
Safety factor	1.57	2.98
Maximum stress (MPa)	319.67	168.63
One-sided reaction force (N)	1007	1521
Amplification ratio by finite element analysis	10.07	15.21
Size: L × W × H (mm^3)	56 × 42 × 29	56 × 42 × 29

Zhou & Zuo, 2013). In addition, a notch-based bridge-type frame (with notch hinges) was presented in Wang et al. (2017). It reduced the stiffness of the beam and enhanced the force amplification ratio.

Fig. 4.3 illustrates the simulation results of stress distribution for the two structures. The maximum stress occurred at the ends of beams for the notch-based frame. That is, for the notch-based bridge-type amplifier, the bending area was concentrated on the ends. Hence, it is insufficient to experience a large force to prevent damage caused by the presence of a stress concentration effect. It is difficult to meet practical demands because of the small safety factor. In contrast, for a beam-based frame, stress on the beam is distributed more uniformly. Therefore, the beam-based bridge-type force amplifier frame was adopted to ensure better safety in this work.

Moreover, to explore the potential of the frame fully, a beam-based frame with a compound design was proposed to strengthen the output performance. The influence of a number of beams for each of the four arms of the bridge-type amplifier was investigated in a simulation study. In view of the practical manufacturing issue, the gap between two adjacent beams was limited to 1.5 mm. In a simulation study, an input force of 400 N was applied and the corresponding reaction force on the output end was monitored to obtain the force amplification ratio. The simulation results of stress distributions for each single force amplifier with different numbers of beams (i.e., one, two, three, and four) are shown in Fig. 4.4. The results are listed in Table 4.4 for a clear comparison.

By increasing the number of beams, the safety factor increased and the maximum stress was reduced. Nevertheless, the force amplification ratio decreased and the physical footprint size increased. Considering the compact size and performance requirement, the two-beam mechanism

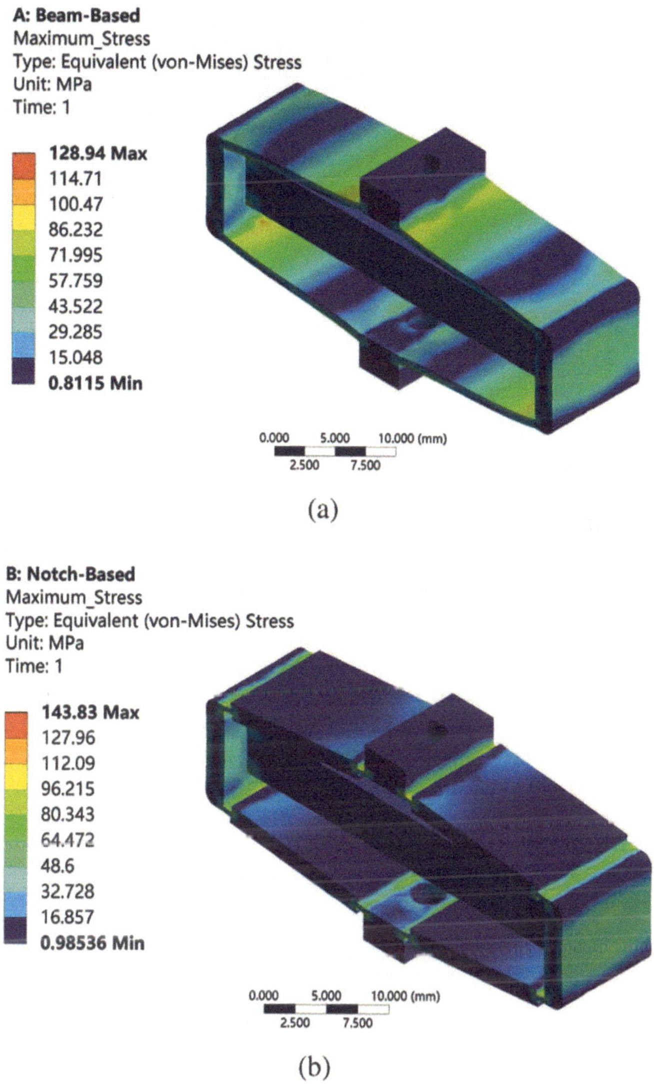

Figure 4.3 Various bridge-type amplification frames. (A) Beam-based frame; (B) notch-based frame.

design was adopted for the compound bridge-type force amplifier in this work, as shown in Fig. 4.4B.

4.3 Analytical modeling

In this section, the analytical model of the designed mechanism is established to predict the force amplification ratio of the force amplifier.

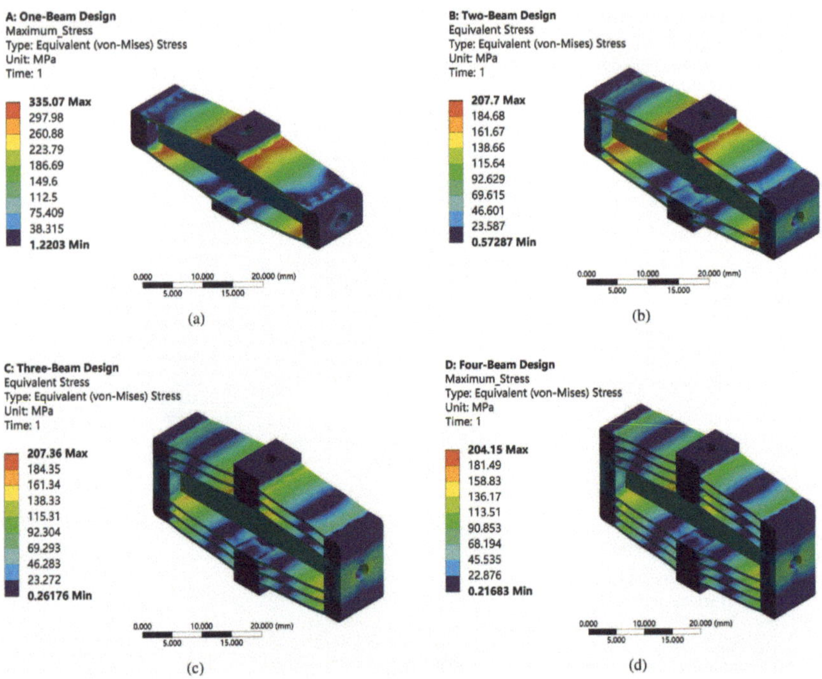

Figure 4.4 Stress distribution of a single compound bridge-type force amplifier with various numbers of beams under an input force of 400 N. (A) One-beam design; (B) two-beam design; (C) three-beam design; (D) four-beam design.

Table 4.4 Performance of single compound bridge-type force amplifier with various numbers of beams for each of four arms under input force of 400 N.

Performance	One-beam design	Two-beam design	Three-beam design	Four-beam design
Safety factor	1.50	2.42	2.89	3.03
Maximum stress (MPa)	335.07	207.70	207.36	204.15
One-sided reaction force (N)	1772	1708	1608	1552
Amplification ratio by finite element analysis	4.43	4.27	4.02	3.88
Size: L × W × H (mm^3)	44 × 18 × 10	44 × 22 × 10	44 × 26 × 10	44 × 30 × 10

Concerning the force amplification ratio of the single-stage force amplification frame, various approaches were reported (Löffler et al., 2017). Most research offers a simplified model by ignoring the elastic properties of the material. However, such a model is able to provide only the relationship between design variables. It cannot accurately evaluate the force amplification ratio. Chen et al. (2017) presented a nonlinear parametric model to calculate the force amplification ratio of the concave type of single-stage force amplification frame, with high accuracy. For a two-stage amplification frame, the analytical model was extensively developed for a frame that functions as a displacement amplifier. However, few theoretical models have been developed for a frame that acts as a force amplifier. In the following discussion, an analytical model is presented to evaluate the convex type of compound two-stage force amplifier.

First, in view of the double symmetry structure, as shown in Fig. 4.5A, one-quarter model of a single amplifier frame is selected for detailed study. As depicted in Fig. 4.5C and D, the compound beam is simplified into an equivalent beam, $A_{12}B_{12}$. The mechanical analysis of the frame is illustrated in Fig. 4.5B–D. The width and thickness of the beam are denoted by b and t, respectively. For the convenience of manufacturing and analytical modeling, the compound frame is designed with the identical value of thickness.

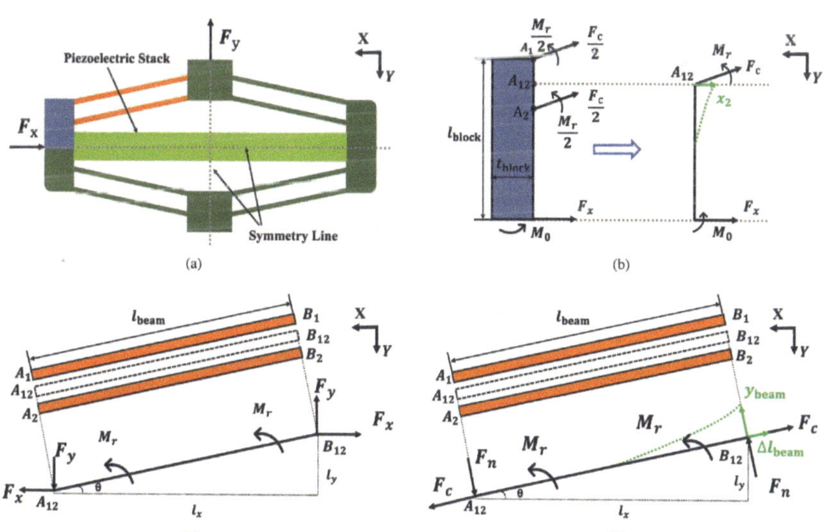

Figure 4.5 Mechanical analysis of a single force amplification frame. (A) The operational principle; (B) side connecting block; (C) the beam; (D) the beam under an equivalent load.

Fig. 4.5C and D show that vertical input force F_y produces horizontal output force F_x. Then, normal force F_n and compressive force F_c along the beam direction can be calculated by equations:

$$F_n = F_y \cos\theta - F_x \sin\theta \tag{4.1}$$

$$F_c = F_y \sin\theta + F_x \cos\theta \tag{4.2}$$

where θ is the tilt angle of the beam with respect to the horizontal direction.

Let l_{beam} and M_r represent the length and applied moment of beam $A_{12}B_{12}$, respectively. Considering the moment equilibrium at point A_{12}, the equation can be derived:

$$2M_r = F_x l_y - F_y l_x = F_x l_{\text{beam}} \sin\theta - F_y l_{\text{beam}} \cos\theta \tag{4.3}$$

$$M_r = \frac{F_x l_{\text{beam}} \sin\theta - F_y l_{\text{beam}} \cos\theta}{2} = \frac{-l_{\text{beam}} F_n}{2} \tag{4.4}$$

In view of the Euler-Bernoulli beam theory, deformation y_{beam} of beam $A_{12}B_{12}$ can be calculated as:

$$\frac{d^2 y_{\text{beam}}}{dl^2} = \frac{M(l)}{E_{\text{frame}} I_{\text{beam}}} \tag{4.5}$$

where E_{frame} denotes the Young's modulus of the frame material, and $I_{\text{beam}} = bt^3/12$ is the area moment inertia of the beam. In addition, $M(l)$ represents the moment at the end point B_{12}, which can be written as:

$$M(l) = F_n l + M_r, \quad 0 \le l \le l_{\text{beam}} \tag{4.6}$$

Referring to Fig. 4.5D, considering the fixed boundary condition at A_{12}, deflection y_{beam} at B_{12} can be obtained below by substituting Eq. (4.6) into Eq. (4.5) and integrating Eq. (1.7) with boundary condition $0 \le l \le l_{\text{beam}}$:

$$y_{\text{beam}}(l_{\text{beam}}) = \frac{F_n l_{\text{beam}}^3}{6E_{\text{frame}} I_{\text{beam}}} + \frac{M_r l_{\text{beam}}^2}{2E_{\text{frame}} I_{\text{beam}}} \tag{4.7}$$

By simplifying Eq. (4.7), the deformation of the beam can be further written into the form:

$$y_{\text{beam}}(l_{\text{beam}}) = \frac{-F_n l_{\text{beam}}^3}{12E_{\text{frame}} I_{\text{beam}}} \tag{4.8}$$

Similarly, let l_{block} and I_{block} represent the half-length and area moment inertia of the side block, respectively. The deformation of the side block for the frame can be obtained:

$$x_2(l_{\text{block}}) = \frac{F_x l_{\text{block}}^3}{3E_{\text{frame}} I_{\text{block}}} \tag{4.9}$$

in which $I_{\text{block}} = b_{\text{block}} t^3{}_{\text{block}}/12$ with b_{block} and t_{block} denote the width and thickness of the side block, respectively.

In addition, the beam is compressed while the input force is exerted on the frame. The elongation Δl_{beam} of the beam becomes:

$$\Delta l_{\text{beam}} = \frac{F_c l_{\text{beam}}^2}{E_{\text{frame}} A_{\text{beam}}} \tag{4.10}$$

where A_{beam} denotes the cross-sectional area of the beam.

Besides, deflection Δx_{pzt} of the piezoelectric stack can be calculated as:

$$\Delta x_{\text{pzt}} = \frac{F_x l_{\text{pzt}}^2}{E_{\text{pzt}} A_{\text{pzt}}} \tag{4.11}$$

Owing to the theorem of compatibility condition, total deformation along the X direction is zero. Therefore, the relationship can be obtained:

$$y_{\text{beam}} \sin\theta = \Delta x_{\text{pzt}} + \Delta l_{\text{beam}} + x_2 \cos\theta \tag{4.12}$$

Substituting Eqs. (4.8)–(4.11) into Eq. (4.12), Eq. (4.12) only contains two variables, F_x and F_y. As a consequence, the force amplification ratio of a single amplifier frame can be derived by $N = Fx/Fy$.

Regarding the compound two-stage force amplification frame, the displacement input and force input of the inner frame are restricted by the connected outer frame. Because of the influence between the two frames, a force transmission coefficient, η, is taken into account. Therefore, the total force amplification ratio of the two-stage force amplification frame can be derived as:

$$N = \eta N_1 N_2 \tag{4.13}$$

where N_1 and N_2 represent the force amplification ratios for the outer and inner frames, respectively.

4.3.1 Prediction of output power

Referring to the literature (Li et al., 2011), an electric model of a piezoelectric stack was developed (Fig. 4.6). The homogenization and

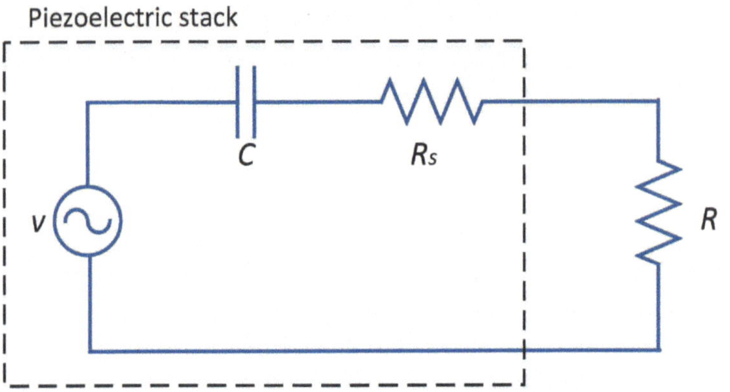

Figure 4.6 Equivalent electric model of a piezoelectric stack.

characterization of a piezoelectric stack actuator was previously investigated (Weiß et al., 2016). The coupled electromechanical modeling of a single convex type of force amplification frame was given by Feenstra et al. (2008), Wang et al. (2016) and Wen and Xu (2017):

$$M\ddot{y}(t) + Ky(t) - hKd_{33}v(t) = F_y(t) \tag{4.14}$$

$$v(t) + hKRd_{33}\dot{y}(t) + RC\dot{v}(t) = 0S \tag{4.15}$$

where t is the time variable, y is the axial displacement, v is the induced voltage, d_{33} is the piezoelectric coefficient, C is the piezoelectric capacitance, and F_y is the input force. In addition, h, M, and K denote the number of layers, mass, and stiffness of the piezoelectric stack, respectively. Also, R is the load resistance. The source impedance of the piezoelectric stack R_s is calculated by:

$$R_s = \frac{1}{j\omega C} \tag{4.16}$$

where ω is the frequency of the excitation signal.

Then, the maximum power output of the harvester P_{max} can be calculated when load resistance R matches the source impedance. That is:

$$P_{\text{max}} = \frac{v_{\text{max}}^2}{R} \tag{4.17}$$

4.4 Dimension optimization and finite element analysis simulation study

4.4.1 Dimension optimization

The dimension optimization of a force amplification frame was conducted (Chen et al., 2017; Wang et al., 2017; Wen & Xu, 2017). The output performance was mostly affected by the tilt angle and thickness of the beam. In practice, given the maximum height of the harvester, the tilt angle can be varied by adjusting the beam length.

As for the output performance, safety and the amplification ratio are the most important aspects, because they influence the capacity of energy harvesting. Moreover, restrictions should be considered when harvesters are applied in practical situations. In particular, the deformation of the beam should be limited to a certain range to minimize unwanted impacts on walking people. In addition, the thickness of the beam should be larger than 0.5 mm, because a thinner beam easily distorts during manufacturing.

As mentioned in the Introduction, the electric output of a piezoelectric stack is proportional to the squared value of load force acting on the piezoelectric stack (Xu et al., 2013). For this reason, the goal of the force amplification mechanism is to magnify force as high as possible to obtain more power output from the piezoelectric stack. Therefore, maximizing force also maximizes the energy output. Related work has also been conducted in the literature (Chen et al., 2017; Wang et al., 2016; Wen & Xu, 2017; Zhou & Zuo, 2013). Hence, the force amplification ratio is the key output performance of the mechanism design. With these settings, optimization is conducted by using the multi-objective genetic algorithm:

- Objective: to maximize the force amplification ratio
- Design variables: the length and thickness of beams
- Subject to:
  (1) The deflection of beam $d < 2$ mm;
  (2) Amplification ratio $N > 21$;
  (3) Safety factor $K > 2$.

Based on this analytical model, the influences of the thickness and length of the beam are obtained, as shown in Fig. 4.7A–D. For the outer frame, a thinner and longer beam exhibits a higher force amplification ratio and lower safety factor. The influence of the thickness is decreased owing to the compound design of the beams. In addition, the safety factor increases when the bending area is enlarged. Therefore, a length of 24 mm and thickness of 0.6 mm are selected to achieve better output performance.

For the inner frame, the output performance is less sensitive to a change in the beam length. Hence, a moderate value of 15 mm is considered. The amplification ratio drops significantly when the thickness is larger than 0.7 mm. In addition, a peak value occurs at a thickness of 0.5 mm. Because of the influence of the design variables, the two values of thickness have a

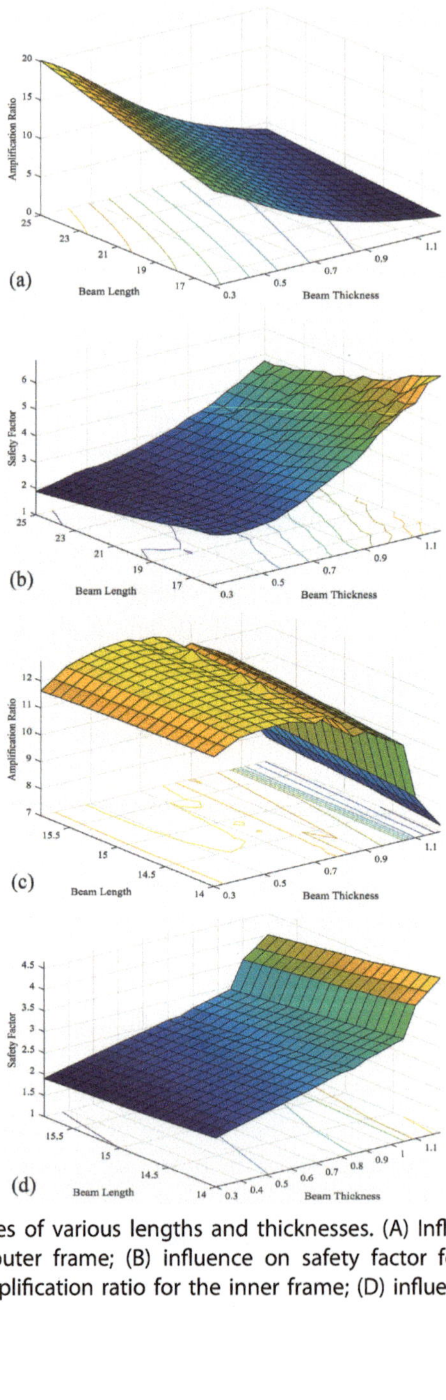

Figure 4.7 Influences of various lengths and thicknesses. (A) Influence on amplification ratio for the outer frame; (B) influence on safety factor for the outer frame; (C) influence on amplification ratio for the inner frame; (D) influence on safety factor for the inner frame.

Table 4.5 Optimal dimension of compound two-stage force amplification frame.

Design variable	Outer frame	Inner frame	Unit
Length	24	15	mm
Width	10	10	mm
Thickness	0.6	0.6	mm
Height	4.9	3.1	mm

similar effect on safety. To ensure high safety and reduce the difficulty of manufacturing, the thickness is chosen to be 0.6 mm rather than the extreme value of 0.5 mm. Moreover, dimension optimization can be extended to include other parameters, such as the width and height. Finally, the optimal variables are assigned as given in Table 4.5.

4.4.2 Finite element analysis simulation results

To verify the output performance, an FEA simulation study was carried out for a conventional single-beam two-stage frame and the proposed compound two-stage frame.

To facilitate the simulation analysis, a force of 100 N was applied on the input end. The reaction force on the output end was generated by the simulation study. The force amplification ratio of the frame was defined as the ratio between the output force and input force. In addition, a stress analysis was carried out to verify the load-bearing capacity of each frame. The simulation results are shown in Figs. 4.8 and 4.9. The performances of the frames are listed in Table 4.6, which shows that a thinner and longer beam produces a larger force amplification ratio for the two-stage frame. Moreover, the safety of the compound two-stage frame is better than that of the two-stage frame without sacrificing too much of the amplification ratio.

The force amplification ratio of each single frame is verified under an input force of 100 N. The simulation results are shown in Table 4.7. The simulation results are consistent with those of the theoretical model. In addition, the force transmission coefficient was calculated to be 0.85 according to the FEA simulation study. The force amplification ratio of the proposed compound frame was better than those in previous work (Zhou & Zuo, 2013).

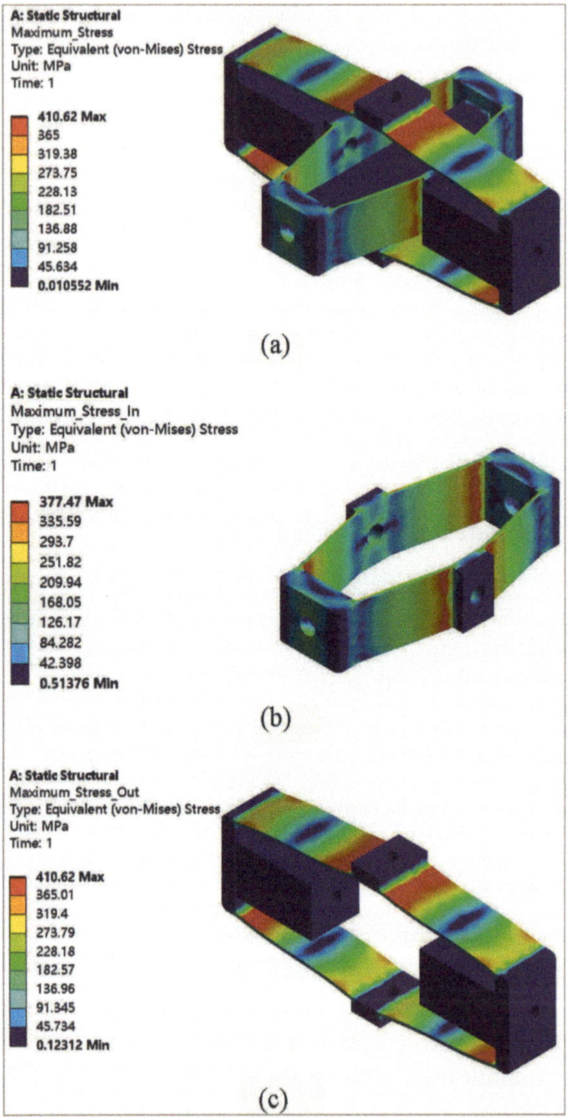

Figure 4.8 Stress distributions of two-stage frame. (A) two-stage frame; (B) inner frame of two-stage frame; (C) outer frame of two-stage frame.

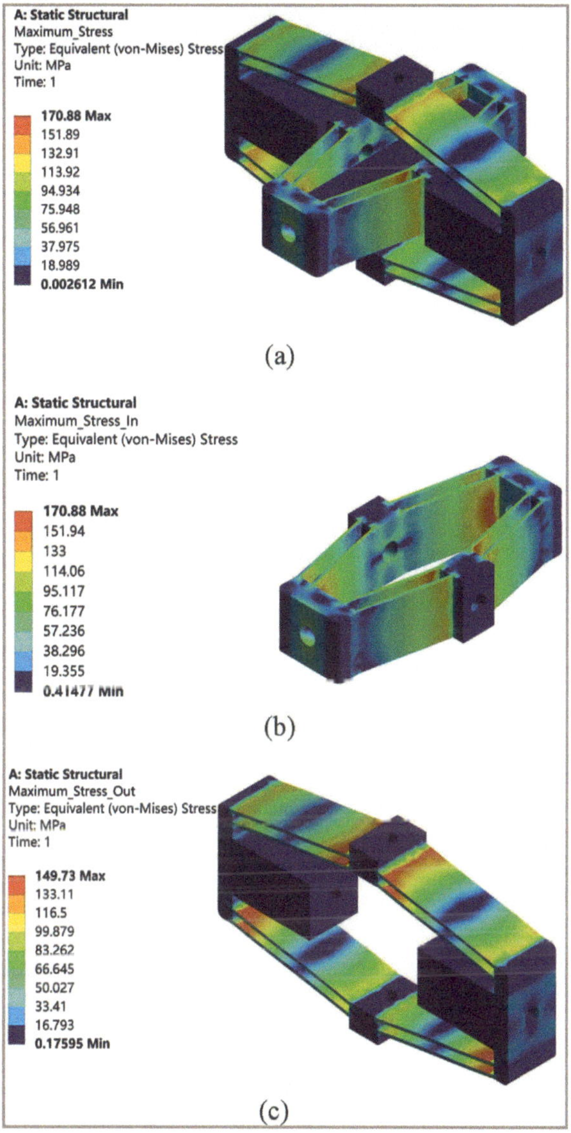

Figure 4.9 Stress distributions of compound two-stage frame. (A) Compound two-stage frame; (B) inner frame of compound two-stage; (C) outer frame of compound two-stage.

Table 4.6 Performance of two-stage frame and compound two-stage frame compared with other two-stage frame under input force of 100 N.

Performance	Single-beam two-stage frame	Compound two-stage frame	Two-stage frame (Wang et al., 2017)
Beam deflection (mm)	4.21	1.31	–
Safety factor	1.23	2.94	1.6
Maximum stress (MPa)	410.62	170.88	1000
One-sided reaction force (N)	2262	1790	–
Amplification ratio by finite element analysis	22.62	17.90	21
Amplification ratio by experiment	–	17.50	21
Size: L × W × H (mm^3)	60 × 44 × 34	60 × 44 × 34	–

Table 4.7 Performance of inner and outer frames of two-stage frame harvester compared with other single-stage frame under input force of 100 N.

Performance	Inner frame (compound two-stage)	Outer frame (compound two-stage)	Single-stage frame (Zhou & Zuo, 2013)
Safety factor	2.94	3.36	–
Maximum stress (MPa)	170.88	149.73	182
One-sided reaction force (N)	433	474	–
Amplification ratio by finite element analysis	4.33	4.74	4.25
Amplification ratio by modeling	4.41	4.77	–
Size: L × W × H (mm^3)	44 × 22 × 10	44 × 34 × 10	–

4.5 Experimental results

4.5.1 Prototype fabrication

A prototype of the proposed compound two-stage force amplification frame was developed to test performance. It was fabricated of a wire electrical discharge machining with optimal parameters. The prototype and experimental setup are shown in Fig. 4.10. It was tested to generate the relationship between the power output and force. Periodic force was applied on the frame using a shaker (Energy Harvesting Kit V1.3, Smart Materials Corp.). Two laser displacement sensors (LK-H055, Keyence Inc.) were used to measure the deflection of the beam compared with the displacement difference between the frame and shaker. A force sensor

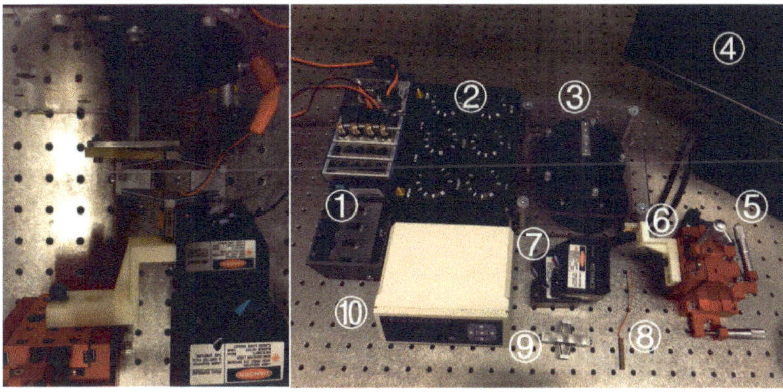

Figure 4.10 Experimental setup of compound two-stage amplification frame. (1) Real-time controller, (2) resistance box, (3) shaker, (4) computer, (5) fixing platform, (6) force sensor, (7) laser displacement sensor, (8) piezoelectric stack, (9) compound two-stage force amplification frame, and (10) signal conditioner of shaker.

(LSB200, FUTEK Advanced Sensor Technology, Inc.) was employed to measure force applied on the input end. A real-time controller (cRIO-9022, National Instruments, Inc.) was used to acquire data. The input impedance of the adopted NI-9215 measurement module was 1 GΩ. In addition, a variable resistance box (ZX21A, Shanghai Precision Instruments, Inc.) was used to examine the maximum power output of the harvester.

4.5.2 Experimental testing results

The experimental results of power output from the compound two-stage harvester and piezoelectric stack under various resistive loads are plotted in Fig. 4.11A and B. The corresponding force and displacement input signals are shown in Fig. 4.12A and B, respectively. The same periodic motion was provided by the shaker to investigate the influence of resistance. For capacitive behavior, the impedance of the piezoelectric stack was calculated as (Wang et al., 2017; Xu et al., 2013):

$$Z_{stack} = \frac{1}{j\omega C_{stack}} \tag{4.18}$$

where ω is the frequency of the excitation signal, and C_{stack} is the capacitance of the piezoelectric stack material.

For electrical impedance matching, maximum power occurs when the resistive load matches the source impedance. Under an excitation frequency

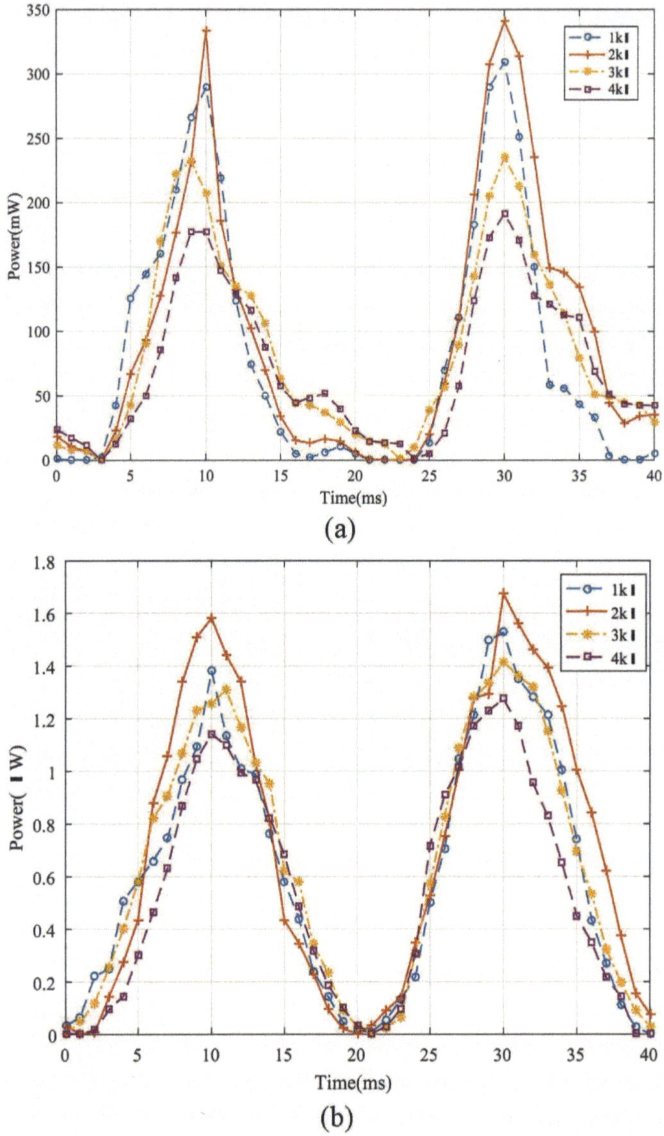

Figure 4.11 Power output with various load resistances at 27 Hz. (A) Compound two-stage harvester; (B) piezoelectric stack.

of 25 Hz, the optimal resistance load is calculated to be 2054 Ω. Then, according to the experimental result shown in Fig. 4.11B, the maximum power of the piezoelectric stack is measured as 1.68 μW with a resistive load of 2000 Ω. Moreover, in Fig. 4.11A, the maximum power of the compound

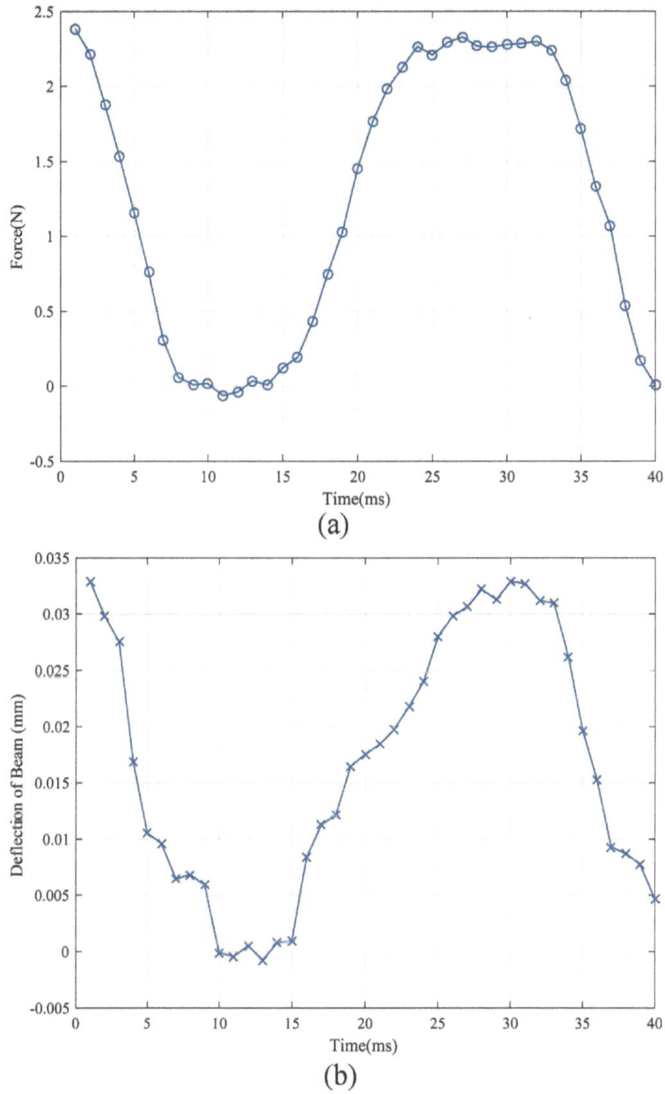

Figure 4.12 Input signals of (A) force and (B) displacement.

two-stage energy harvester at a resistive load of 2000 Ω is 340.7 μW, which is 203 times more than the performance of the piezoelectric stack. In this test, the maximum force input is 2.33 N. Hence, the capacity of energy harvesting is derived as 130.77 μW/N. For further improvement, a piezoelectric stack with smaller electrical capacities can be used to produce a higher ratio between the output voltage and applied force.

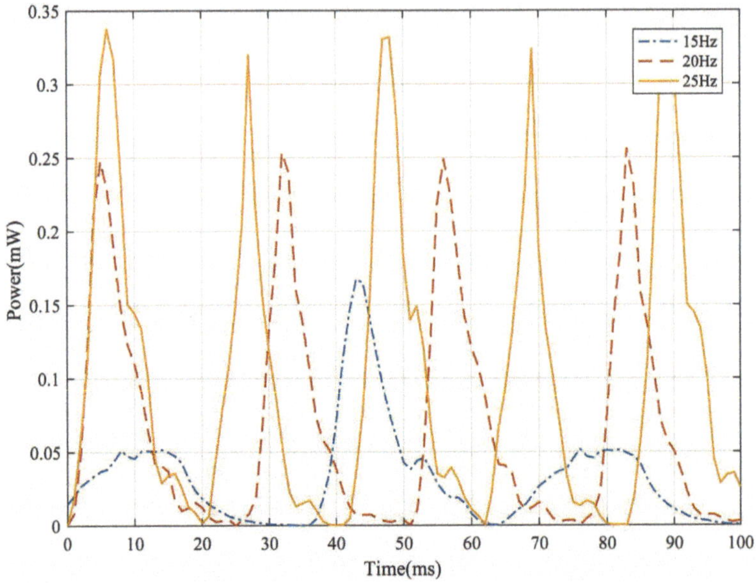

Figure 4.13 Influence of various external frequencies on the compound two-stage frame.

Similarly, the output performance of the harvester was tested under different conditions of excitation frequency. The measurement setup is shown in Fig. 4.10 and the results are depicted in Fig. 4.13. The power output improved as the excitation frequency increased under the same load resistance. The optimal excitation frequency of the compound two-stage harvester occurred at 25 Hz across a resistor of 2000 Ω. Because of the influence of periodic force, the motion of the compound two-stage harvester is not as smooth as typical periodic motion, especially around the crests and valleys. This effect could be caused by the imperfect rigid connection between the force sensor and the harvester.

Concerning the influence of the excitation frequency, the voltage output of the piezoelectric stack is not always consistent with variations in the input force. When the input force suddenly increased to a specific value and maintained that value, the voltage output of the compound two-stage harvester dropped (drifted) slowly compared with the input force, as shown in Fig. 4.14.

The amplification ratio was tested by generating the relationship between the voltage and force. As shown in Fig. 4.15, curve-fitting models

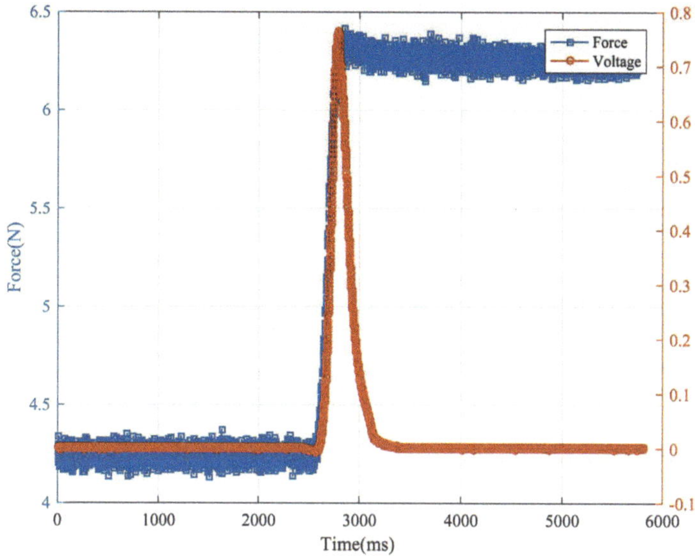

Figure 4.14 Effect of sudden increase, keeping the external load.

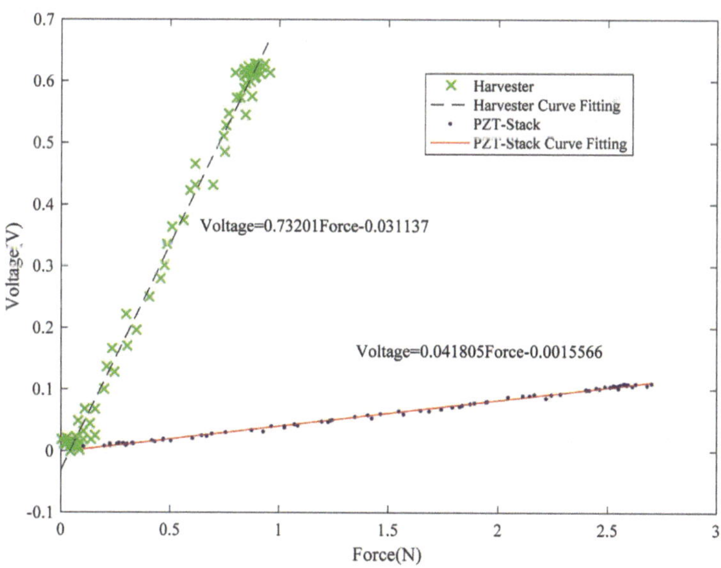

Figure 4.15 Experimental testing results of amplification ratio for the compound two-stage amplification frame.

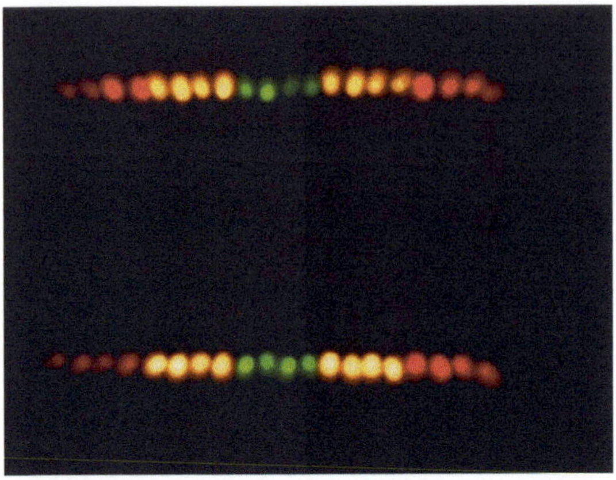

Figure 4.16 Power output testing results with 40 LEDs 3 mm in diameter connected in parallel.

were obtained according to the voltage output under different forces for the harvester and piezoelectric stack. The amplification ratio was equal to the slope of the fitted linear curve. The slope of the harvester's fitted curve was 17.5 times that of the piezo-stack's fitted curve. The experimental result was close to the simulation result of 17.9. The discrepancy was induced by manufacturing tolerance and assembly error. The difference also occurred because the two frames are considered to be a whole system in the FEA simulation test. In practice, they are connected by screws, which is not as rigid as that assumed in the simulation study.

As a case study of practical application, the compound two-stage harvester was used to light at least 40 LEDs. The LEDs had a diameter of 3 mm and were connected in parallel, as shown in Fig. 4.16. Experimental results demonstrated the feasibility of the proposed compound two-stage harvester to power LEDs. In addition, because of the compact size of the harvester, it can be adopted to design a tile with multiple (e.g., nine) compound two-stage energy harvesters of identical specification. Therefore, the proposed harvester has a wide range of use for more promising applications.

4.6 Conclusions

This chapter presents the design and development of a piezoelectric energy harvester with a compound two-stage force amplification frame. Compared with other force amplification frames, it is able to bear a larger external load and provide a higher amplification ratio. To predict the amplification ratio of the frame accurately, an analytical model was established based on the Euler-Bernoulli beam bending theory. Dimension optimization was carried out to obtain the maximal output performance. A prototype harvester was fabricated and the performance was verified by performing both FEA simulation and experimental studies. The relationships between the input force and output power were almost linear. The power output of the piezoelectric stack is related to the mechanism design of the frame and also to the external load resistance and excitation frequency. For future work, a platform of a compound two-stage harvester with a more sophisticated energy harvesting circuit will be developed to scavenge more energy from human footsteps.

References

Chen, W., Wang, Y., & Deng, W. (2017). Deformable force amplification frame promoting piezoelectric stack energy harvesting: Parametric model, experiments and energy analysis. *Journal of Intelligent Material Systems and Structures, 28*(7), 827–836. https://doi.org/10.1177/1045389X16657426

Feenstra, J., Granstrom, J., & Sodano, H. (2008). Energy harvesting through a backpack employing a mechanically amplified piezoelectric stack. *Mechanical Systems and Signal Processing, 22*(3), 721–734. https://doi.org/10.1016/j.ymssp.2007.09.015

Li, X., Guo, M., & Dong, S. (2011). A flex-compressive-mode piezoelectric transducer for mechanical vibration/strain energy harvesting. *IEEE Transactions on Ultrasonics, Ferroelectrics, and Frequency Control, 58*(4), 698–703. https://doi.org/10.1109/TUFFC.2011.1862

Löffler, M., Weiß, M., Wiesgickl, T., & Rupitsch, S. J. (2017). Untersuchung analytischer und numerischer Modelle zur anwendungsspezifischen Dimensionierung eines Amplified Piezo Actuators. *Technisches Messen, 84*(11), 706–718. https://doi.org/10.1515/teme-2017-0028

Wang, Y., Chen, W., & Guzman, P. (2016). Piezoelectric stack energy harvesting with a force amplification frame: Modeling and experiment. *Journal of Intelligent Material Systems and Structures, 27*(17), 2324–2332. https://doi.org/10.1177/1045389X16629568

Wang, L., Chen, S., Zhou, W., Xu, T. B., & Zuo, L. (2017). Piezoelectric vibration energy harvester with two-stage force amplification. *Journal of Intelligent Material Systems and Structures, 28*(9), 1175–1187. https://doi.org/10.1177/1045389X16667551

Weiß, M., Rupitsch, S. J., & Lerch, R. (2016). Homogenization and characterization of piezoelectric stack actuators by means of the inverse method. In *Proc. of 2016 joint IEEE international symposium on the applications of ferroelectrics, European conference on application of polar dielectrics, and piezoelectric force microscopy workshop (ISAF/ECAPD/PFM)* (pp. 1–4).

Wen, S., & Xu, Q. (2017). Design of a two-stage force amplification frame for piezoelectric energy harvesting. In *Proc. of IEEE international conference on cybernetics and intelligent systems (CIS) and IEEE conference on robotics, automation and mechatronics (RAM)* (pp. 1–6).

Xu, T. B., Siochi, E. J., Kang, J. H., Zuo, L., Zhou, W., Tang, X., & Jiang, X. (2013). Energy harvesting using a PZT ceramic multilayer stack. *Smart Materials and Structures, 22*(6). https://doi.org/10.1088/0964-1726/22/6/065015

Zhou, W., & Zuo, L. (2013). A novel piezoelectric multilayer stack energy harvester with force amplification. In *Proc. of ASME 2013 international design engineering technical conferences and computers and information in engineering conference* (pp. 1–8).

CHAPTER 5

Design of a piezoelectric energy harvesting handrail with dual excitation modes

This chapter presents the design, fabrication, and testing of an energy harvesting handrail. The handrail energy harvester has two working modes for collecting energy excited by both vibration and a pulling force during the operation. To enable good stability and antijamming for the energy harvester, a compound bridge-type compliant force amplifier is designed and optimized. An analytical dynamic model is established for the piezoelectric device and the output power is calculated. A prototype harvester is fabricated for experimental investigation. Experimental results verify the effectiveness of the derived analytical model. A series of designed energy harvesters can be used to scavenge energy generated by passengers in a bus or subway under different motion states.

5.1 Introduction

With the rapid development of the Internet of Things, more and more miniaturized and low-power electronic products have entered into our daily lives. Compared with traditional battery-powered supply methods, the use of energy scavenged from the ambient environment has emerged as a convenient and promising way to power miniaturized and low-power electronic devices (Shirvanimoghaddam et al., 2019). Public buses and subways are the main approach to urban transportation; they provide sufficient energy sources for harvesting. Some energy harvesters have been reported to collect vibration energy from moving vehicles (Wiwiek et al., 2012; Xie & Wang, 2015). As a tool for transporting passengers, the bus or subway offers vibration energy and provides human movement energy. In this chapter, a handrail energy harvester is developed to scavenge energy from vibration and force excitations in public vehicles.

Mechanical Design of Piezoelectric Energy Harvesters
ISBN 978-0-12-823364-1
https://doi.org/10.1016/B978-0-12-823364-1.00001-8
© 2022 Elsevier Inc.
All rights reserved.

In this chapter, an energy harvesting handrail with two working modes is designed and developed that can collect vibration energy when a bus in driven and from the pulling force energy produced by the passenger. The proposed energy harvesting handrail is composed of a bridge-type force amplifier with an embedded piezoelectric stack and an outer protection frame. In experiments, an acceleration sensor was used to acquire the movement characteristics of the handrail when the bus was driven. According to its characteristics, the parameters of the bridge-type force amplifier were optimized to achieve the most efficient energy transmission. Analytical modeling, a simulation study, and experimental verification were carried out to demonstrate the effectiveness of the proposed energy harvesting handrail.

5.2 Mechanism design

Using the public bus as an example of an application scenario, the actual working environment of the handrail energy harvester is introduced in this section (Fig. 5.1). Then, an appropriate bridge-type force amplifier is designed and its parameters are optimized. In addition, to make the mechanism work safely, an external protection mechanism is designed for the harvester.

5.2.1 Working principle

A computer-aided design (CAD) model of the proposed energy harvesting handrail is shown in Fig. 5.2. The handrail energy harvester is composed of

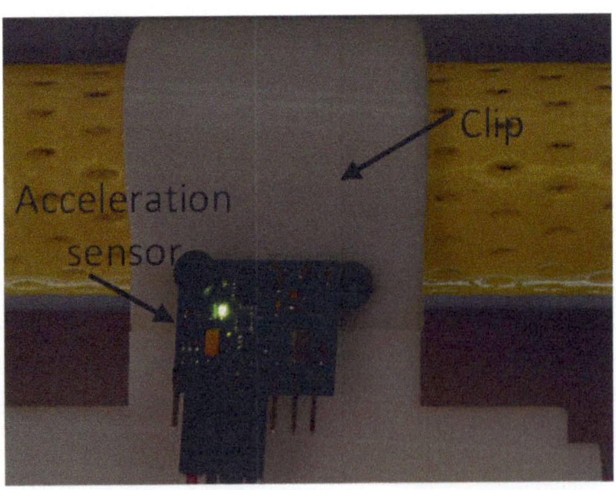

Figure 5.1 Movement data acquisition of a bus handrail with an acceleration sensor.

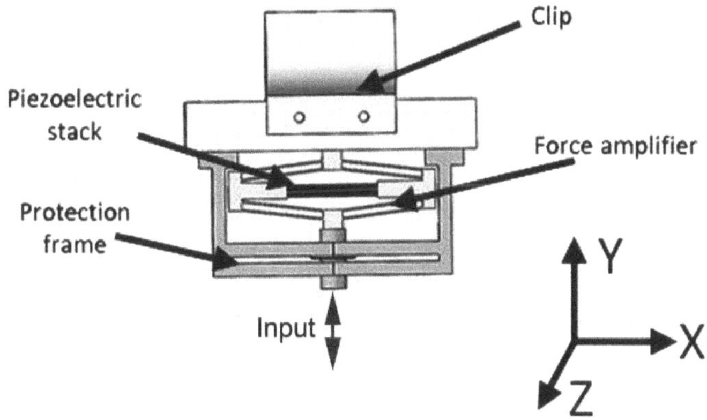

Figure 5.2 Computer-aided design model of the energy harvesting handrail.

a piezoelectric stack, a force amplifier mechanism, a protection frame, and a clip for fixing onto the bus. The parameters of the force amplifier and protection frame are shown in Fig. 5.3. When an inertial force (owing to vibration) or pulling force (by the passenger) is applied on the input end, the force will be amplified by the force amplifier mechanism and then transmitted onto the two ends of the piezoelectric stack. The piezoelectric stack will generate electric power based on the direct piezoelectric effect.

Before the mechanism of the harvester is designed, data on the movement of a bus handrail are acquired by an acceleration sensor when the bus is driven. The collected data are shown in Fig. 5.4, which indicate that the handrail has two types of motion states. One is a small-amplitude, high-frequency vibration and the other is a large-amplitude, low-frequency vibration, as depicted in Figs. 5.5A and 5.6A, respectively. Moreover, a spectrum analysis of the collected time-domain acceleration signal is conducted using fast Fourier transform. The single-sided amplitude spectrums are illustrated in Figs. 5.5B and 5.6B, respectively.

The small-amplitude, high-frequency vibrations represent the normally driven bus, whereas the large-amplitude, low-frequency vibrations indicate that the bus starts and stops (e.g., arriving and leaving the station) or turns. When the bus is started, stopped, and turned, a large pulling force is exerted on the handrail by the passenger owing to the inertial effect. At this moment, the impact of the pulling force on the handrail is much greater than that of the vibration.

Because of actual working conditions, the energy harvesting handrail needs to possess two working modes. One collects energy excited by the

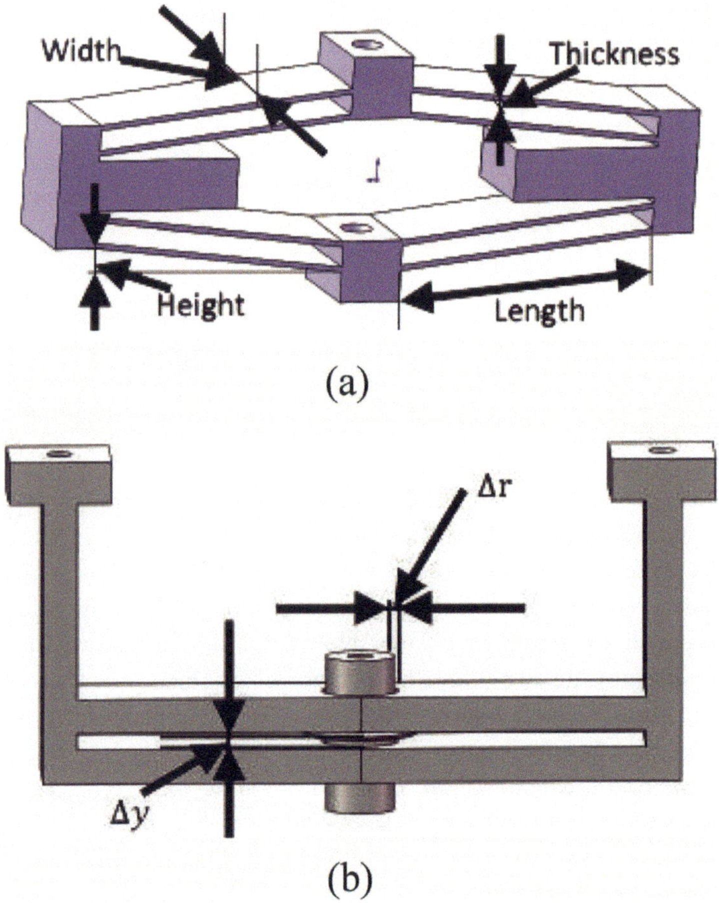

Figure 5.3 Representation of parameters for (A) force amplifier and (B) protection frame.

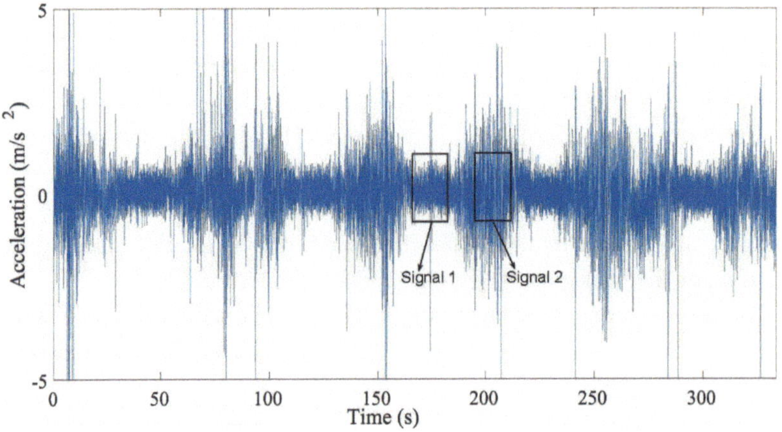

Figure 5.4 Acquired acceleration signal data of the bus handrail.

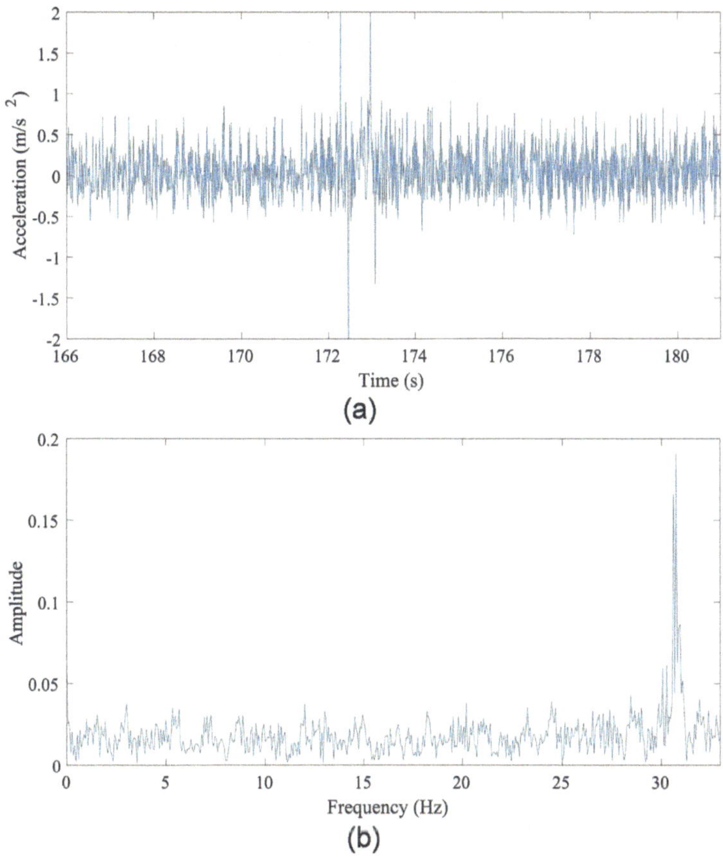

Figure 5.5 High-frequency vibration signal. (A) Signal with small-amplitude, high-frequency vibration; (B) single-sided amplitude spectrum of the high-frequency vibration signal.

small-amplitude, high-frequency vibration of the bus and the other scavenges energy produced by a pulling force on the handrail, which is exerted by the passenger.

5.2.1.1 Force amplifier mechanism

A CAD model of the force amplifier mechanism is shown in Fig. 5.3A, which is a compound bridge-type compliant force amplifier. The force amplifier is a core component of the energy harvesting handrail, which can improve the energy conversion efficiency of the piezoelectric stack. The workspace of the energy harvesting handrail is in three dimensions; there is vibration excitation in the Y direction as well as interference in the X and Z

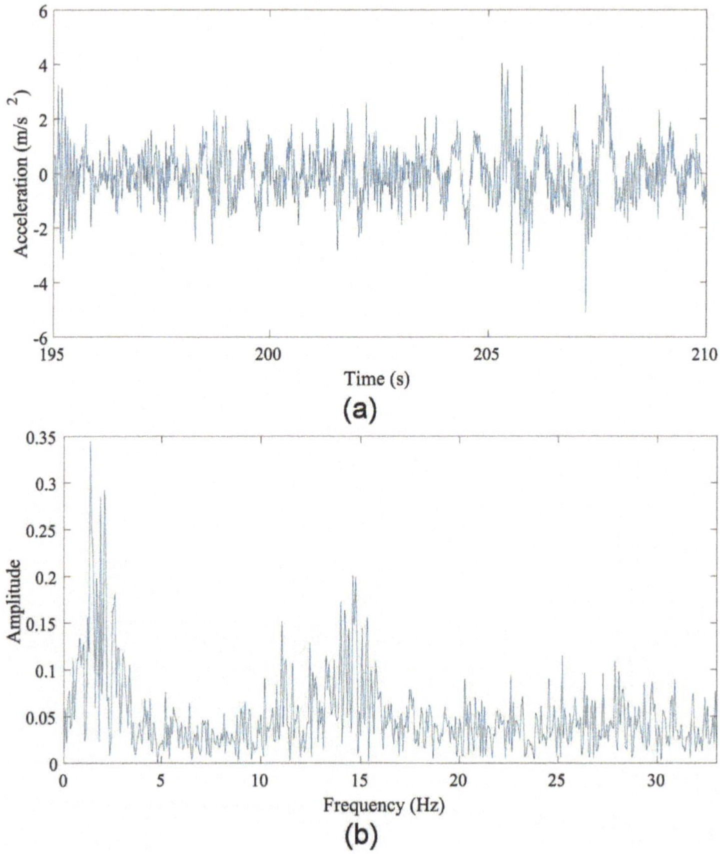

Figure 5.6 Low-frequency vibration signal. (A) Signal with large-amplitude, low-frequency vibration; (B) single-sided amplitude spectrum of the low-frequency vibration signal.

directions (especially in the X direction, when the bus has stopped or started). The compound bridge-type force amplifier has larger lateral stiffness and better antijamming ability than an ordinary bridge-type amplifier, and it is more suitable for a working environment (Xu & Li, 2011). This is a reason for choosing a compound bridge-type force amplifier in this work.

5.2.1.2 Protection frame mechanism

To prevent material damage to the mechanism (owing to excessive tensile stress) and limit the impact of deformation on the passenger, a protection frame is designed, as illustrated in Fig. 5.3B. It consists of three parts: the left and right brackets and an intermediate connection cylinder. The two brackets and the cylinder are connected to the bus fixing mechanism and force amplifier mechanism by screws. When the vertical deformation of the force amplifier mechanism reaches Δy, the wafer contacts the bracket and prevents further deformation, protecting the mechanism.

Another function of the protection frame is to reduce the interference of forces on the mechanism. The ideal situation is that the mechanism only experiences a vertical downward pulling force. In practice, however, a pulling force is applied at a random angle to the vertical direction. The horizontal component of the pulling force may induce damage to the mechanism. Thus, it is necessary to offset the component of the pulling force at certain level, which is constrained by radial clearance Δr. When the horizontal displacement reaches Δr, the cylinder will contact the bracket and prevent further horizontal movement, protecting the mechanism. In this work, the vertical and horizontal displacement ranges are limited by $\Delta y = 1$ mm and $\Delta r = 0.5$ mm, respectively.

5.2.2 Parameter optimization of force amplifier mechanism

The output performance of the force amplifier is primarily governed by flexure beam dimensions such as length, height, thickness, and width. Therefore, these dimensions are selected to be the key design parameters. To optimize the performance of the force amplifier mechanism, the key parameters are optimally designed by resorting to the multiobjective genetic algorithm (MOGA) based on finite-element analysis (with ANSYS Workbench software). The material of the amplifier is selected to be Al-7075, whose main properties are shown in Table 5.1.

First, to achieve a higher voltage and power output, the dominant objective of optimization is to maximize the force amplification ratio of the force amplifier. The safety factor is defined as the ratio between the allowable maximum stress and actual maximum stress of the material, and

Table 5.1 Main parameters of material Al-7075.

Parameter	Value
Density	2810 kg/m^3
Poisson's ratio	0.33
Young's modulus	26.96 GPa
Shear modulus	71.7 GPa
Tensile yield strength	503 MPa

Table 5.2 Optimization results of flexure beam parameters.

Parameter	Value (mm)
Length	25
Height	3
Thickness	0.5
Width	10

the safety factor should be greater than 1 in practice. Thus, to prevent material failure, the safety factor of the force amplifier needs to be greater than 2.0 under the two working modes. In addition, to minimize the unwanted impact on the passenger's comfort, it is necessary to limit deformation of the force amplifier to less than 1 mm (Δy).

Based on these constraints, the parameters of the force amplifier mechanism are optimized as shown in Table 5.2. When the force amplifier is preloaded with 20-N pulling force and given a vibration excitation with an amplitude of 0.01 mm and frequency of 30.5 Hz (the same as in the experimental study), the simulation results for the total deformation, the safety factor, and equivalent elastic strain of the mechanism are shown in Fig. 5.7A–C, respectively. When a pulling force of 300 N is applied to the force amplifier, the simulation results for total deformation, the safety factor, and equivalent elastic strain are depicted in Fig. 5.8A–C, respectively. Under an input force of 300 N, the induced maximal deformation and minimal safety factor are 1.005 mm and 2.316, respectively. In addition, the input force of 300 N produces an output force of 2394 N by the force amplifier, which is exerted on the piezoelectric stack. Thus, the force amplification ratio of the mechanism is approximately 8.

5.3 Modeling of piezoelectric stack

A piezoelectric stack is adopted as the energy conversion material based on the d_{33} effect. There are many methods for modeling the piezoelectric stack; the single-degree-of-freedom dynamic model has been adopted because of its good precision and easy implementation (Goldfarb & Celanovic, 1997; Qian et al., 2018a; Zhu et al., 2014). Based on this approach, the electromechanical coupling effect of the piezoelectric stack is formulated, and the voltage and power outputs of the energy harvesting handrail are predicted.

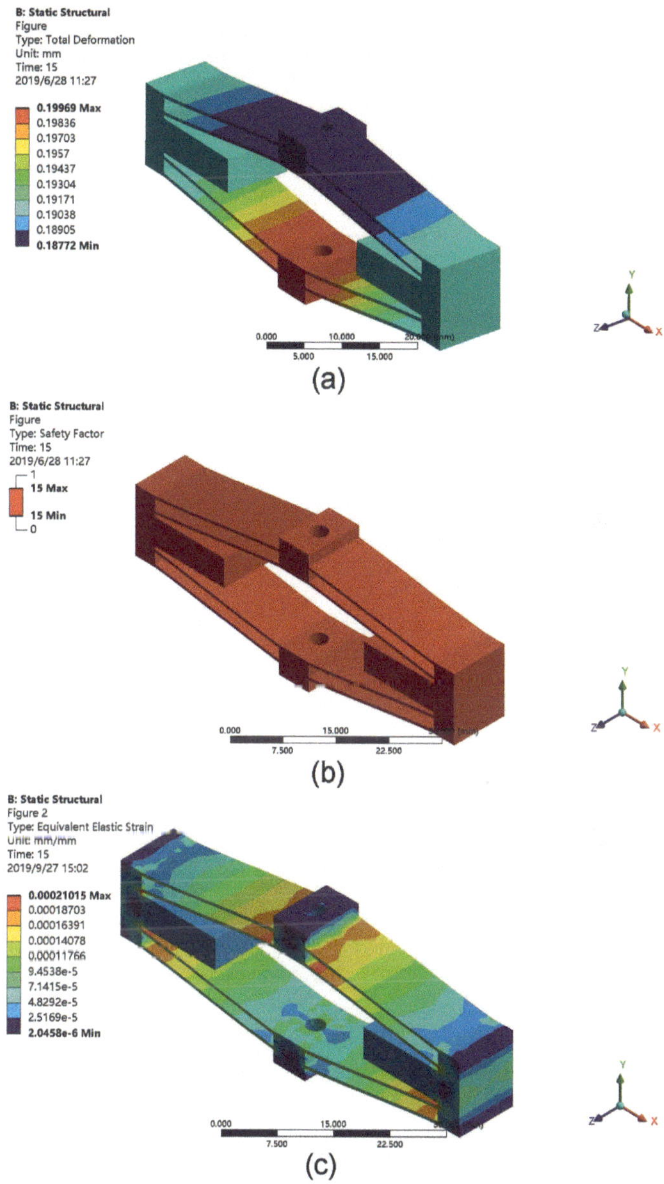

Figure 5.7 Finite-element analysis simulation results. (A) Total deformation, (B) safety factor, and (C) equivalent elastic strain of force amplifier mechanism under 20-N pulling force along with vibration excitation.

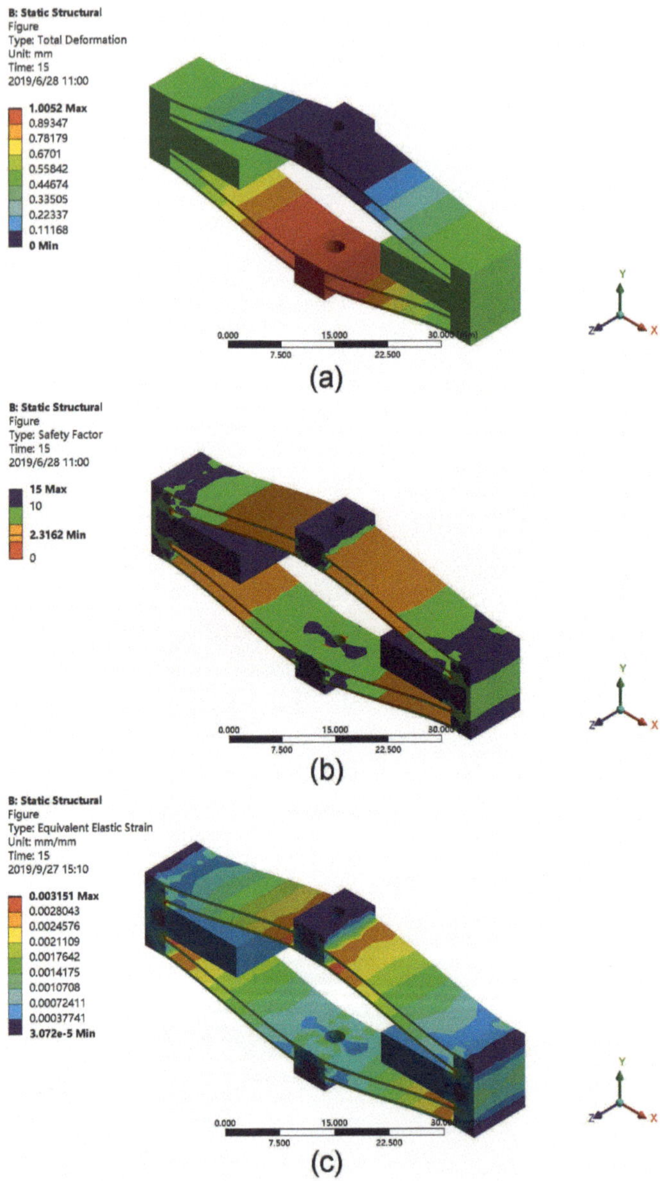

Figure 5.8 Finite-element analysis simulation results. (A) Total deformation, (B) safety factor, and (C) equivalent elastic strain of force amplifier mechanism under 300-N pulling force.

The detailed structure of the multilayer piezoelectric stack in parallel connection is shown in Fig. 5.9A, where t_p is the thickness of a single piezoelectric layer, n is the number of layers of the piezoelectric stack, L is the length of the piezoelectric stack, and $v(t)$ is the voltage output of the piezoelectric stack under the action of an external force. According to the actual situation, the dynamic model of the piezoelectric stack can be simplified to represent the axial vibration of a flexible rod under pressure applied at both ends, as illustrated in Fig. 5.9B, where $F_p(t)$ is the piezoelectric force, $F(t)$ is the force applied to the piezoelectric stack, m is the equivalent mass of the piezoelectric stack, and $x(t)$ is the displacement. Damping of the piezoelectric stack has little influence on the dynamic characteristics and is negligible. Thus, the dynamic equation of the single-degree-of-freedom model is established by Newton's second law (Feenstra et al., 2008; Qian et al., 2018b):

$$m\ddot{x}(t) = F(t) - F_p(t) \tag{5.1}$$

To derive the relationship between applied force $F(t)$ and output voltage $v(t)$, it is necessary to understand the electromechanical coupling effect of the piezoelectric material. The linear constitutive equations of piezoelectric stack can be written as (Feenstra et al., 2008; Qian et al., 2018b):

$$S_3(t) = s_{33}^E T_3(t) + d_{33} E_3(t) \tag{5.2}$$

$$D_3(t) = d_{33} T_3(t) + \varepsilon_{33}^T E_3(t) \tag{5.3}$$

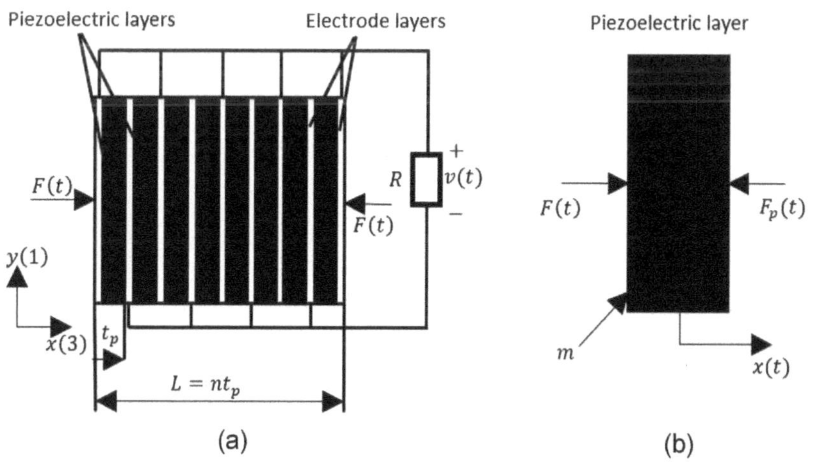

Figure 5.9 Illustration of (A) schematic diagram of the piezoelectric stack in parallel connection; (B) electromechanical coupling model of the piezoelectric stack.

Eqs. (5.2) and (5.3) represent the mechanical response and electrical response, respectively, of the piezoelectric material. The axial strain and stress of the piezoelectric stack under applied force $F(t)$ are denoted by $S_3(t)$ and $T_3(t)$, respectively. $E_3(t)$ and $D_3(t)$ are the electric field density and electric displacement, respectively. s_{33} is the compliance, d_{33} is the piezoelectric coupling coefficient, and ε_{33} is the dielectric permittivity. The subscripts denote the direction of each property. Superscript E indicates that parameter s_{33} is measured at a constant electric field. Superscript T indicates that parameter ε_{33} is measured at constant stress. Because the piezoelectric films are connected in parallel, the current working frequency is smaller than the natural frequency of the force amplification frame with an inner piezoelectric stack. The strain, stress, electric field density, and electric displacement can be related to displacement $x(t)$, force $F(t)$, voltage $v(t)$, charges over all cross-sectional areas $q(t)$, and cross-sectional area A. The relationships are written as:

$$S_3(t) = \frac{x(t)}{L},\ T_3(t) = \frac{F(t)}{A},\ E_3(t) = \frac{V(t)}{t_p},\ D_3(t) = \frac{q(t)}{A} \tag{5.4}$$

Substituting these terms into Eqs. (5.2) and (5.3) gives:

$$x(t) = \frac{L}{AE}F_p(t) + \frac{Ld_{33}}{t_p}v(t) \tag{5.5}$$

$$q(t) = nd_{33}F(t) + \frac{n\varepsilon_{33}^{T}A}{t_p}v(t) \tag{5.6}$$

where $E = \frac{1}{s_{33}^{E}}$ is the elastic Young's modulus. The expression of $F_p(t)$ can be obtained by solving Eq. (5.5). Then, $F_p(t)$ is substituted into dynamic Eq. (5.1), which results in the expression:

$$m\ddot{x}(t) + \frac{EA}{L}x(t) - \frac{EAd_{33}}{t_p}v(t) = F(t) \tag{5.7}$$

In the absence of an external electric field, and by neglecting the influence of damping, the elastic deformation of the piezoelectric stack under an external force can be approximated by a spring with the longitudinal stiffness of $k = EA/L$. The relationship between applied force $F(t)$ and displacement $x(t)$ is described by $F(t) = kx(t)$. Substituting this relationship into the electrical response in Eq. (5.6) yields:

$$q(t) = nd_{33}kx(t) + \frac{n\varepsilon_{33}^{T}A}{t_p}v(t) \tag{5.8}$$

As shown in Fig. 5.9A, the external resistance is R and the voltage is $v(t)$. Based on Ohm's law, the result is that:

$$-\frac{v(t)}{R} = \dot{q}(t) = nd_{33}k\dot{x}(t) + C_p\dot{v}(t) \tag{5.9}$$

where $q(t)$ is the circuit current, R denotes resistance, and $C_p = n\varepsilon_{33}TAt_p$ is the capacitance of the piezoelectric stack. Further simplifying Eq. (5.9) and combining it with Eq. (5.7) gives:

$$m\ddot{x}(t) + kx(t) - nkd_{33}v(t) = F(t) \tag{5.10}$$

$$\dot{v}(t) + \frac{v(t)}{C_pR} + \frac{nd_{33}k}{C_p}\dot{x}(t) = 0 \tag{5.11}$$

Once the force applied on the piezoelectric stack is known, the output voltage of the piezoelectric stack can be predicted by solving those equations. Furthermore, the output power can be calculated by:

$$P(t) = \frac{v^2(t)}{R} \tag{5.12}$$

5.4 Prototype fabrication and experimental testing results

In this work, a prototype of the energy harvesting handrail is fabricated with optimal parameters, and the harvester's performance is verified through experimental study.

5.4.1 Prototype fabrication

A fabricated prototype of the energy harvesting handrail is shown in Fig. 5.10. The compound bridge-type force amplifier is fabricated by wire-electrical discharge machining technology with Al-7075 material. The protection frame is made of steel material. The fixing clip is fabricated by a three-dimensional printer with ABSplus material. The experimental setup is shown in Fig. 5.10B. A shaker (KSI-758ST150, Beijing KingSci Instruments, Inc.) provides vibration excitation to the harvester. A triaxial displacement platform provides support and applies a preloading force to the mechanism. A force sensor (LSB200, Futek Advanced Sensor Technology, Inc.) and a laser displacement sensor (LK-H055, Keyence Inc.) are respectively used to measure applied force and vibration displacement for

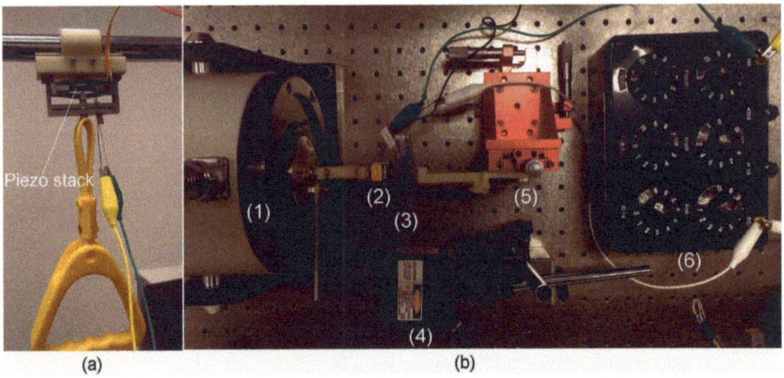

Figure 5.10 (A) Prototype and (B) experimental setup of energy harvester. (1) Shaker, (2) force sensor, (3) compound bridge-type force amplifier, (4) laser displacement sensor, (5) triaxial displacement platform, and (6) resistance box.

the mechanism. A variable resistance box (ZX21A, Shanghai Precision Instruments, Inc.) provides an external resistor for the experimental study. The measured data are collected by a data acquisition device (PXIe-1082, National Instruments, Inc.). The main parameters of the piezoelectric stack (RP150/5x5/30, Harbin Soluble Core Tech Co., Ltd.) are shown in Table 5.3.

5.4.2 Test results with vibration excitation

First, a vibration test was conducted on the energy harvesting handrail to verify the feasibility and effectiveness of the proposed structural design.

The vibration signals of the bus handrail collected by an acceleration sensor are shown in Fig. 5.5A. A spectrum analysis was performed and the results are shown in Fig. 5.5B. They indicate that the vibration frequency

Table 5.3 Main parameters of piezoelectric stack.

Prototype	Value
Elastic modulus (E)	54 GPa
Piezoelectric coupling (d_{33})	650 pC/N
Relative permittivity ($\varepsilon^T{}_{33}/\varepsilon_0$)	2400
Capacitance (C)	3.1 μF
Cross-sectional areas	25 mm^2
Length	28 mm
Mass	4.0 g
Number of layers	230

and amplitude of the handrail acceleration were approximately 30.5 Hz and 0.46 m/s^2, respectively. To facilitate the experimental study, this work adopted a sinusoidal wave with the same frequency and amplitude to approximate real working conditions of the handrail. The displacement characteristics of the handrail, whose amplitude and frequency are 0.01 mm and 30.5 Hz, respectively, were obtained through double integration. The displacement vibration was used as an excitation for the mechanism vibration test. When the bus was driven, the passenger applied a pulling force on the handrail, which was measured at about 20 N. Therefore, a 20-N preloading force needed to be applied to the mechanism when the vibration test was performed.

Eq. (5.12) shows that the output power is inversely proportional to the external resistance. Thus, to obtain optimal power output, it is necessary to select a suitable resistor. The variation in average power output for the energy harvesting handrail with various resistive loads is plotted in Fig. 5.11. Because of the capacitance behavior of the piezoelectric stack, impedance can be calculated as:

$$Z_{\text{stack}} = \frac{1}{j\omega C_p} \tag{5.13}$$

where $\omega = 2\pi \times 30.5$ rad/s is the frequency of the excitation signal. Based on electrical impedance matching, maximum power output occurs when the resistive load matches the source impedance. According to Eq. (5.13), optimal resistance is calculated as $1683\ \Omega$. Then, according to the

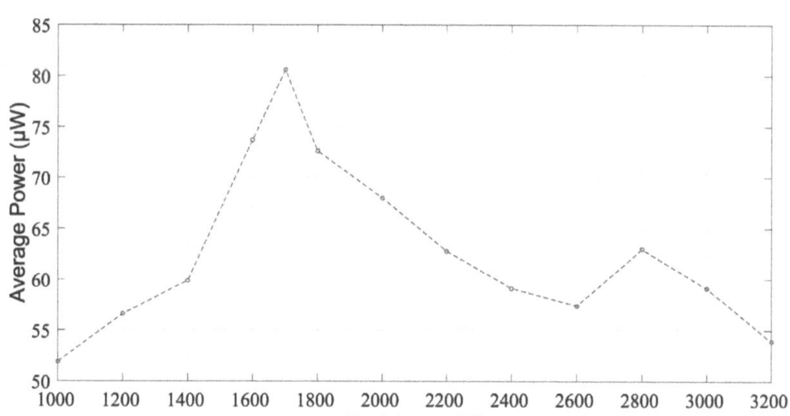

Figure 5.11 Experimental results of power output under vibration excitation with different load resistances.

experimental result in Fig. 5.11, the maximum average power output of the energy harvesting handrail is 80.6 μW when the resistive load is 1700 Ω. This indicates that the experimental result is consistent with the calculated value of resistance, with a small discrepancy of only 1%.

The energy harvesting handrail was tested several times; a typical result for the vibration test is shown in Fig. 5.12. The calculated and measured voltage and power outputs of the energy harvesting handrail at a resistive load of 1700 Ω are given in Fig. 5.12A and B, respectively. Under vibration excitation, the peak-to-peak voltage output of the harvester was about 1 V, and the amplitude of the output power for the harvester was about 150 μW. Fig. 5.12 shows that the output voltage and power of the energy harvester

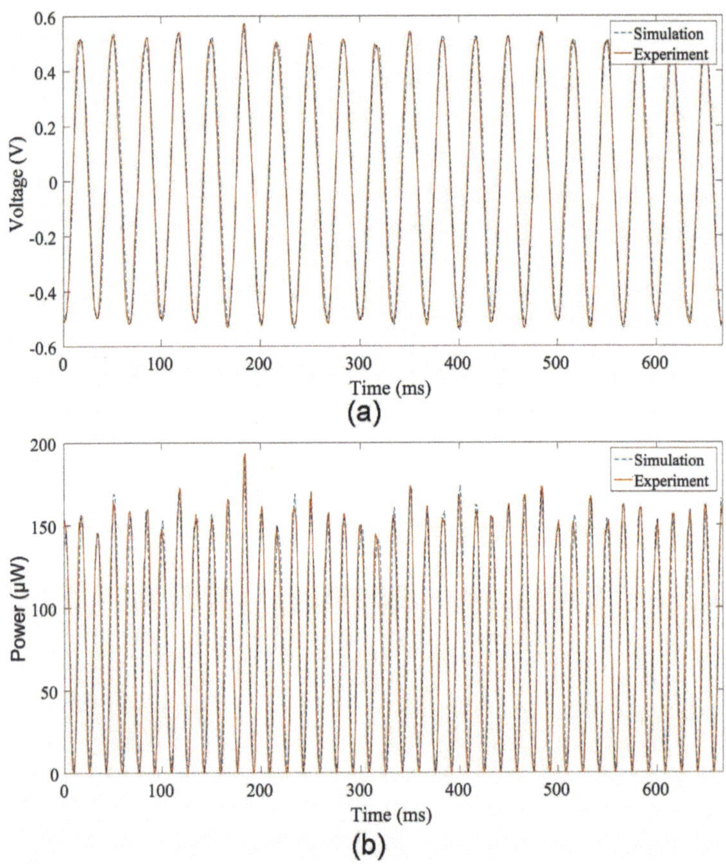

Figure 5.12 Experimental and simulation results of voltage and power output of energy harvester under vibration excitation. (A) Voltage output; (B) power output.

obtained by simulation and experiment were close. The average powers generated by experiment and simulation are 80.6 and 80.3 μW, respectively, with a discrepancy of 0.37%. The experimental and simulation results show that the energy harvester can effectively collect vibration energy when the bus is driven. Although the scavenged energy is relatively low, only about 80 μW, it is enough to supply low-power electronic devices.

5.4.3 Test results with random pulling force

Second, the energy harvesting handrail was tested by applying a pulling force. For the situations in which the bus stops, starts, and turns, the acceleration characteristics of the handrail and its frequency analysis are shown in Fig. 5.6A and B, respectively. At the moment of such motion states, the passenger exerts a large pulling force on the handrail owing to the inertial effect.

A schematic diagram of the random pulling test for the handrail is shown in Fig. 5.10. The harvester was fixed with a cross-beam through a clip. The clip, protection frame, and force amplifier were fixed by screws. The performance of the harvester was tested by randomly pulling on the handrail. Moreover, to measure the pulling force, two strain gauges with a half-bridge connection were used to construct a force sensor to measure the pulling force applied on the handrail.

According to the frequency distribution characteristics of low-frequency modes as shown in Fig. 5.6B, the dominant frequency was about 1.5 Hz. According to Eq. (5.13), a theoretical optimal resistance of 34,227 Ω is calculated at 1.5 Hz. Therefore, 35,000 Ω is chosen as the external resistance for the random tension test. Under a pulling force input given in Fig. 5.13A, the simulated and measured voltage and power outputs of the energy harvesting handrail at a resistive load of 35,000 Ω are illustrated in Fig. 5.13B and C, respectively. The peak voltage output of the harvester with a random pulling test is more than 20 V. The peak output power of the harvester under a random pulling test is about 15 mW. In addition, as shown in Fig. 5.13, under a random pulling force, the output voltages and powers of the energy harvester obtained by simulation and measurement are close to each other. Compared with the vibration excitation result, the deviation between the simulation and experimental results of the random pulling test is relatively larger. The maximum power difference reaches 9.8%. The experimental and simulation results reveal that the energy harvester designed in this work can effectively collect energy induced by a pulling force when the bus is driven.

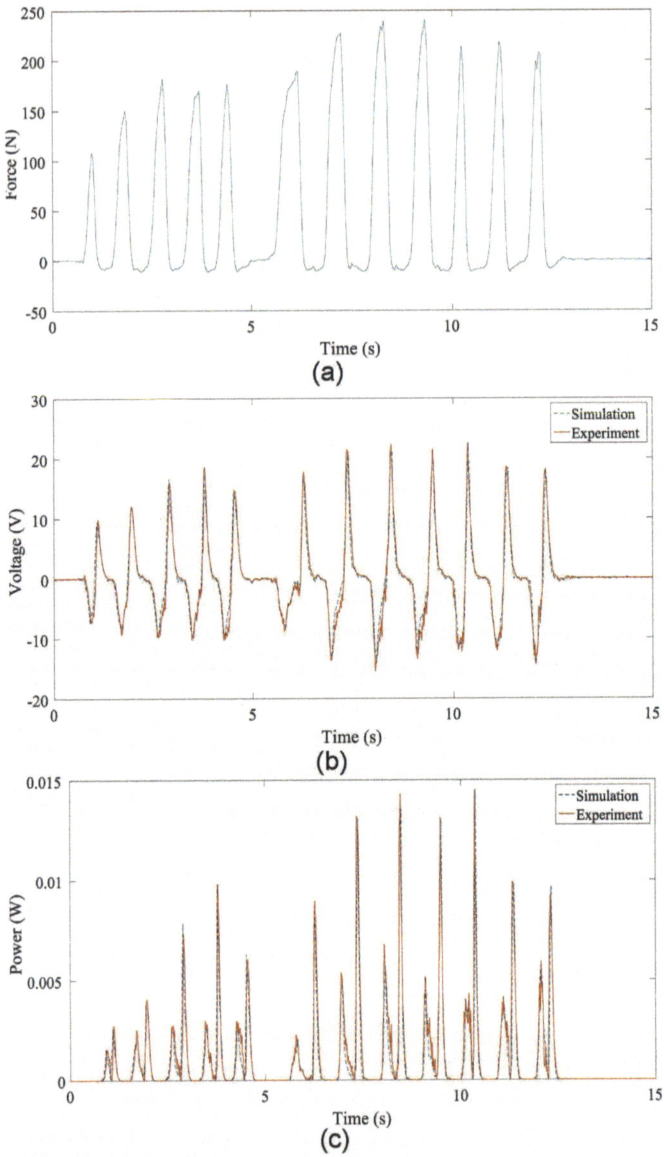

Figure 5.13 Experimental and simulation results of voltage and power output of energy harvester under random pulling input. (A) Random pulling input; (B) voltage output; (C) power output.

5.4.4 Discussion

Through an experimental test of the energy harvesting handrail, it is found that the energy harvester provides a higher power output under a pulling force excitation, and the generated energy is about 100 times that under vibration excitation. This indicates that the piezoelectric stack is more suitable for collecting energy from the pulling force input. However, compared with the piezoelectric patch, the piezoelectric stack has advantages for recovering vibration energy. Using two piezoelectric patches (12 mm × 10 mm × 0.6 mm), Wang et al. (2017) reported an energy output of 30.55 μW at 7 km/h, whose energy density was 0.212 μW/mm^3. In the current work, the power and energy density obtained by the adopted piezoelectric stack (4.0 g) were 150 μW and 0.214 μW/mm^3, respectively. That is, the developed harvester provides a larger power output and exhibits the same level of power density as in the previous work, as shown in Table 5.4. The superiority of the generated power density is not as obvious because of the relatively small vibration amplitude of the application environment in this work.

In addition, the test results for the proposed harvester with a random pulling force were compared with those of the previous work (Feenstra et al., 2008; Wu & Xu, 2019), as listed in Table 5.5. The output power produced by the presented harvester reached 15 mW, which is much larger than that in previous work (Feenstra et al., 2008; Wu & Xu, 2019). In addition, the energy density obtained by the random pulling test in this work was 21.4 μW/mm^3, which is 21 and 116 times those in previous work (Feenstra et al., 2008; Wu & Xu, 2019), respectively. Better performance of the proposed energy harvester is enabled because it can withstand a greater pulling force.

Table 5.4 Comparison of harvester performance with vibration input.

	This work	Reference (Wang et al., 2017)
Power output	150 μW	30.55 μW
Energy density	0.214 μW/mm^3	0.212 μW/mm^3

Table 5.5 Comparison of harvester performance with pulling force input.

	This work	Reference (Feenstra et al., 2008)	Reference (Wu & Xu, 2019)
Power output	15 mW	0.4 mW	128.51 μW
Energy density	21.4 μW/mm^3	1 μW/mm^3	0.184 μW/mm^3

As an example of an application, multiple energy harvesting handrails can be installed in a bus in practice. Hence, considerable energy can be collected by energy harvesting handrails. The scavenged energy is sufficient to power alarm bells and signal lights on the bus. Moreover, the collected energy can be used for low-power electronic devices (such as wireless Bluetooth headsets) for passengers on the bus.

5.5 Conclusion

In this chapter, a type of energy harvesting handrail with two working modes was designed and tested. The handrail consists of a clip, force amplifier, piezoelectric stack, and protection frame to collect energy generated when the bus is driven. The force amplifier was optimally designed using the MOGA method, which provides the largest force amplification ratio and safety factor. A dynamic model was derived to characterize the piezoelectric stack to predict the output voltage and power of the energy harvesting handrail. In both vibration and random pulling tests, the numerical results are in good agreement with the experimental results. This indicates that the developed model can be used to predict the output voltage and power for the energy harvester. Experimental results show that the maximum voltage and power output of the harvester are 1 V and 150 μW under a vibration state, and 20 V and 15 mW under a random pulling state. Both experimental and simulation results reveal that the harvester can be used to collect energy generated by the bus handrail in different motion states. Future work includes improving the design of the harvester mechanism to enhance the output power density further.

References

Feenstra, J., Granstrom, J., & Sodano, H. (2008). Energy harvesting through a backpack employing a mechanically amplified piezoelectric stack. *Mechanical Systems and Signal Processing, 22*(3), 721–734. https://doi.org/10.1016/j.ymssp.2007.09.015

Goldfarb, M., & Celanovic, N. (1997). Modeling piezoelectric stack actuators for control of micromanipulation. *IEEE Control Systems, 17*(3), 69–79. https://doi.org/10.1109/37.588158

Qian, F., Xu, T. B., & zuo, L. (2018a). A distributed parameter model for the piezoelectric stack harvester subjected to general periodic and random excitations. *Engineering Structures, 173*, 191–202. https://doi.org/10.1016/j.engstruct.2018.06.076

Qian, F., Xu, T. B., & Zuo, L. (2018b). Design, optimization, modeling and testing of a piezoelectric footwear energy harvester. *Energy Conversion and Management, 171*, 1352–1364. https://doi.org/10.1016/j.enconman.2018.06.069

Shirvanimoghaddam, M., Shirvanimoghaddam, K., Abolhasani, M. M., Farhangi, M., Zahiri Barsari, V., Liu, H., Dohler, M., & Naebe, M. (2019). Towards a green and self-powered Internet of Things using piezoelectric energy harvesting. *IEEE Access, 7*, 94533–94556. https://doi.org/10.1109/ACCESS.2019.2928523

Wang, W., Cao, J., Bowen, C. R., Zhou, S., & Lin, J. (2017). Optimum resistance analysis and experimental verification of nonlinear piezoelectric energy harvesting from human motions. *Energy, 118*, 221–230. https://doi.org/10.1016/j.energy.2016.12.035

Wiwiek, H., Laksana, G. H., & Nyoman, S. I. (2012). Design, modeling and analysis of implementing a multilayer piezoelectric vibration energy harvesting mechanism in the vehicle suspension. *Engineering*, 728–738. https://doi.org/10.4236/eng.2012.411094

Wu, Z., & Xu, Q. (2019). Design and testing of a novel bidirectional energy harvester with single piezoelectric stack. *Mechanical Systems and Signal Processing, 122*, 139–151. https://doi.org/10.1016/j.ymssp.2018.12.026

Xie, X. D., & Wang, Q. (2015). Energy harvesting from a vehicle suspension system. *Energy, 86*, 385–392. https://doi.org/10.1016/j.energy.2015.04.009

Xu, Q., & Li, Y. (2011). Analytical modeling, optimization and testing of a compound bridge-type compliant displacement amplifier. *Mechanism and Machine Theory, 46*(2), 183–200. https://doi.org/10.1016/j.mechmachtheory.2010.09.007

Zhu, W., Chen, G., Bian, L., & Rui, X. (2014). Transfer matrix method for multibody systems for piezoelectric stack actuators. *Smart Materials and Structures, 23*(9). https://doi.org/10.1088/0964-1726/23/9/095043

CHAPTER 6

Design of a piezoelectric energy harvester based on multistage force amplification frame

This chapter reports a novel integrated multistage (IMS) energy harvester. It is designed to scavenge mechanical energy from human footsteps during walking. By fully exploiting the characteristics of the force amplification frame, the harvester provides the function of position protection, which enables a large load to be borne and thus protects the piezoelectric stack. The maximum displacement input is restricted to a certain range to ensure walking comfort. A static analytical model of the energy harvester is established through force analysis based on the stiffness matrix method. The main architectural parameters of the device are optimized by adopting the multiobjective genetic algorithm (MOGA). A prototype is fabricated for experimental investigation. Experimental results show that the harvester has a large force amplification ratio and high peak power output for practical application.

6.1 Introduction

Because a piezoelectric stack with multilayer piezoelectric material results in high energy density, energy harvesters based on a piezoelectric stack with a force amplification frame have been investigated to increase power output (Yang & Zu, 2016). Electrical energy is generated when external stress is applied on both sides of the piezoelectric stack, which works based on the d_{33} mode of piezoelectric coefficient (Xu et al., 2013). A force amplification frame combined with a piezoelectric stack can magnify exerted stress to maximize power output. Typically, the harvester contains at least one bridge-type force amplifier and a piezoelectric stack, which has a large force amplification ratio and high output performance. Such energy harvesters

Mechanical Design of Piezoelectric Energy Harvesters
ISBN 978-0-12-823364-1
https://doi.org/10.1016/B978-0-12-823364-1.00003-1
© 2022 Elsevier Inc.
All rights reserved.

have been studied extensively (Feenstra et al., 2008; Wang, Chen, et al., 2017; Wang et al., 2016; Wen & Xu, 2018; Zhou & Zuo, 2013). However, most work tried to optimize the performance of force amplifiers while ignoring their applicability. The prevalent bridge-type force amplifier has only one pair of input ends and one pair of output ends, leading to a small contact area for external force. As a result, it is hard to guarantee the safety of the structure with stress exerted on anything other than a compressive direction. Moreover, each force amplifier can bear a limited load, and it may be damaged when the external load exceeds the material's limit. For practical application, a protection mechanism is required to ensure the safety of the harvester. This means that extra structural components are needed to compensate for the loss of safety, resulting in an increase in the size and complexity of the device.

To overcome these issues and obtain higher output performance, various force amplification mechanisms have been presented to remedy the weaknesses of the bridge-type amplifier (Liu & Yan, 2016; Su & Yang, 2002; Wang, Ding, et al., 2017; Xie & Cai, 2014). For instance, a wedge structure is a simple force amplification mechanism that can bear a large force safely. However, its force amplification ratio is relatively small (Xie & Cai, 2014). A leverage structure has a large force amplification ratio; its drawbacks are low safety and a large physical size (Liu & Yan, 2016; Su & Yang, 2002; Wang, Ding, et al., 2017). Because each force amplification mechanism has its own features, it is desirable to design an IMS force amplifier to achieve high output performance and tolerate a high load from the ambient environment. To this end, an energy harvester is devised in this chapter by integrating four different types of force amplification mechanisms. A comparison study of similar designs (Wang, Chen, et al., 2017; Wen et al., 2018) based on the piezoelectric stack reveals that the proposed harvester has better performance.

This chapter presents the design and development of an integrated multistage energy harvester to improve the safety of force amplifiers under a large loading situation. The safety factor is defined as the ratio of the yield stress of the frame's material to actually experienced maximal stress of the energy harvester. Thus, the concept of safety factor is employed to measure the load-carrying capacity of the system in this work. Both simulation and experimental studies are carried out to demonstrate the output performance of the developed energy harvester.

6.2 Mechanical design

A computer-aided design model of the proposed integrated multistage energy harvester is shown in Fig. 6.1. To achieve high output performance under a large force condition, it integrates four types of force amplification frames. It can be classified into three parts: the top plate, inner force amplifier, and bottom plate. As shown in Fig. 6.1C, the inner force amplifier is composed of an output wedge, leverage structure, half-bridge mechanism, and bridge-type force amplifier. As illustrated in Fig. 6.1D, the top and bottom plates consist of an input wedge and side support.

6.2.1 Working principle

The working principle of the proposed energy harvester is explained as follows. When an external force (F_i) is applied on the top plate, the vertical motion of the top plate is converted into the horizontal motion of the output wedge structure. The bottom plate with the same structure is adopted to ensure smooth translational motion and to keep the forces in

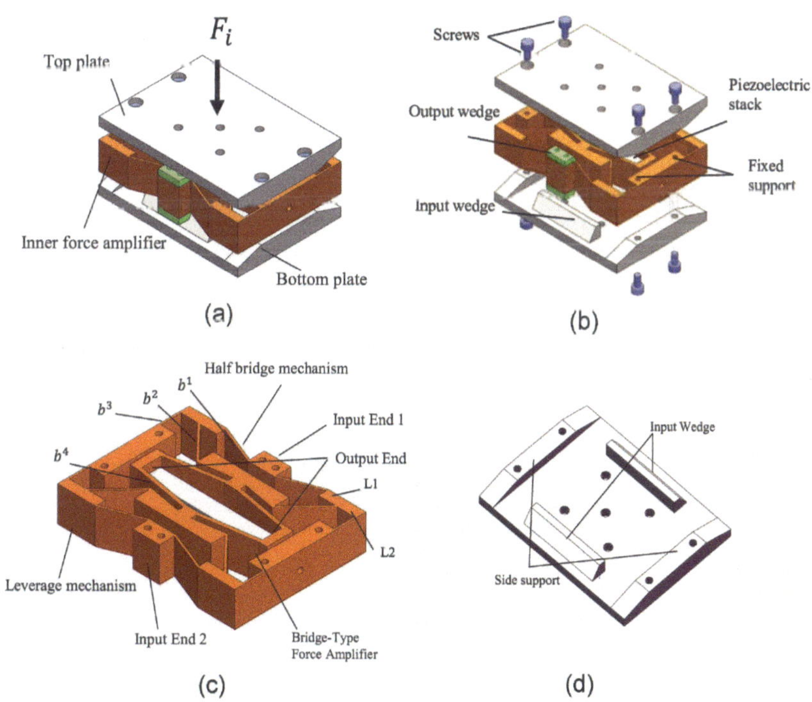

Figure 6.1 Computer-aided design model of designed multistage energy harvester. (A) Isometric view; (B) exploded view; (C) force amplifier; (D) top and bottom plates.

equilibrium. The force is first amplified by the wedge mechanism in this process. The motion of the output wedge structure for the inner force amplifier drives the half-bridge mechanism and leads to rotational motion for each of the four leverages. The parasitic rotational motion of the entire leverage structure is suppressed owing to the constraint of the double symmetric mechanism. In this stage, force is amplified again by the half-bridge and leverage mechanisms. Finally, the bridge-type force amplifier, which is connected to the output end of the leverage mechanism, further magnifies force exerted on the piezoelectric stack, resulting in more power output.

6.2.2 Mechanism design

To have a better understanding of the harvester mechanism, the design of each submechanism is presented in detail.

6.2.2.1 Wedge mechanism

The wedge mechanism is selected as the first-stage force amplification frame owing to its structural characteristics. It can transfer and magnify vertical force into a horizontal force. The top and bottom plates are supported by wedge components, as depicted in Fig. 6.2. The contact area between the input and output wedge structures is large enough to withstand a large load without damaging the material. Moreover, the entire top plate of the harvester is taken as the input area, which differs from a traditional bridge-type force amplifier. Therefore, the stability of the energy harvester is enhanced. The drawback of such a mechanism is that the friction force, which appears on the bevel between two wedge structures, restrains relative motion. This leads to a decrease in the force amplification ratio.

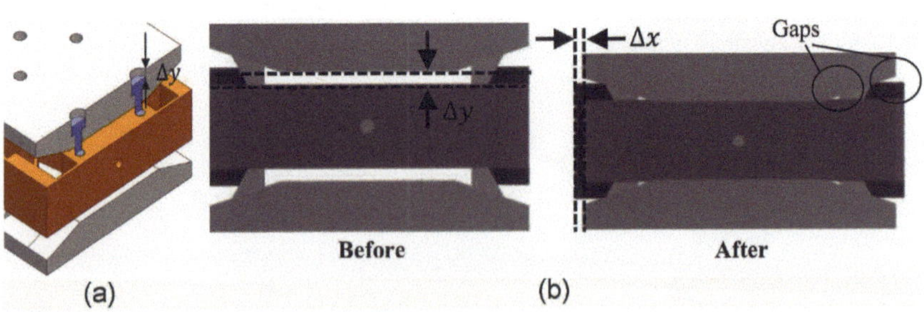

Figure 6.2 Schematic diagram of (A) sectional view and (B) side views before and after applying force.

6.2.2.2 Leverage structure with half-bridge mechanism

The leverage structure includes three parts: the input end, output end, and pivot. The connections among them are implemented by compliant flexure beams. To magnify the input displacement or force, the elastic deformation of the beams leads to mechanical transformation in the leverage structure (Su & Yang, 2002). Furthermore, a leverage structure with a concave type of half-bridge mechanism can enhance the performance of the force amplification and constrain the physical size. In addition, the pivot of the leverage structure can be considered a fixed support to connect extra mechanical components. In addition to minimizing the effect of shear force, the symmetric leverage structure functions as a guiding mechanism to maintain linear motion.

6.2.2.3 Bridge-type force amplifier

The main stage of the harvester is a bridge-type force amplifier. Owing to the requirements of a high safety factor and large force amplification ratio, the convex beam-based force amplification frame is chosen based on previous work (Wen et al., 2018).

6.2.2.4 Position protection mechanism

To improve the stability and safety of the structure, the energy harvester should restrict the input stroke to a certain range. When an external load (F_i) is applied, the energy harvester is compressed until the input stroke limit is reached. Then, the harvester will no longer be compressed regardless of the magnitude of the exerted load. In this design, the input stroke is defined as the sum of gaps between the two plates and the inner force amplifier. With full-range compression, plates with side supports will contact the fixed support of the inner force amplifier, forming a stable supporting structure.

As shown in Fig. 6.2A, it is assumed that the screws have the same height as the plate after it is compressed. Side views of the IMS energy harvester are illustrated in Fig. 6.2B before and after force is exerted. There are two gaps between the inner force amplifier and the two plates to ensure that their motion will not be affected. The force amplification mechanism functions as springs. Once stress is released, the two plates will be restored to their original positions by the inherent elastic force. Moreover, screws are used to fix the two plates with the inner force amplifier, which are mounted within the two plates with a depth equal to the value of the input stroke.

Furthermore, the circular through-holes of the two plates have the same dimension as the screws to restrict movement of the IMS energy harvester anything other than a compressive direction, as shown in Figs. 6.1D and 6.2A. It enhances the stability of the device and improves the comfort of people walking on the plate.

By further exploiting the benefits of the integrated force amplifiers, the harvester has the function of position protection, which ensures that the harvester material will not be damaged under a large load.

6.2.3 Material selection

The energy harvester is fabricated with aluminum 7075-T6 because of its relatively high tensile yield strength and light weight. The dominant mechanical properties of the material are listed in Table 6.1. In addition, because of the requirement of fine d_{33} performance, the piezoelectric stack of model P-885.91 (Physik Instrumente Co., Ltd.) is selected. The main specifications of the adopted piezoelectric stack are listed in Table 6.2.

6.3 Analytical modeling

In this section, analytical static models of the energy harvester are established and architectural parameters are optimized to achieve the best output performance.

Table 6.1 Mechanical properties of aluminum 7075 material.

Property	Value
Poisson's ratio	0.33
Young's modulus	71.7 GPa
Density	2810 kg/m^3
Tensile yield strength	503 MPa

Table 6.2 Electromechanical properties of adopted piezoelectric stack at d_{33} mode.

Property	Symbol	Value
Elastic compliance coefficient	S_{33}^{E}	20.7×10^{-12} m^2/N
Piezoelectric voltage coefficient	g_{33}	25×10^{-3} Vm/N
Piezoelectric charge coefficient	d_{33}	400×10^{-12} C/N
Relative permittivity	$\varepsilon_{33}^{T}/\varepsilon_0$	2400
Coupling factor	k_{33}	0.69
Capacitance	C_p	3.1×10^{-6} F

6.3.1 Model of wedge mechanism

First, the static model for the wedge mechanism is established by analyzing the geometric relationship between the wedge components, as shown in Fig. 6.3. For the input wedge, a relationship can be obtained by considering that the net force of the system is zero:

$$N_1 \sin \alpha - \mu_1 N_1 \cos \alpha = N_0 \tag{6.1}$$

$$N_1 \cos \alpha + \mu_1 N_1 \sin \alpha = F_i \tag{6.2}$$

where N_1, α, and μ_1 are the normal force, tilt angle, and friction coefficient of the bevel, respectively; N_0 is the reaction force for the sliding support; and F_i is the force input.

Similarly, the governing equation for the output wedge is given by:

$$N_1 \sin \alpha - \mu_1 N_1 \cos \alpha = F_o \tag{6.3}$$

$$N_1 \cos \alpha + \mu_1 N_1 \sin \alpha = N_2 \tag{6.4}$$

where N_2 is the reaction force of sliding support and F_o is the output force.

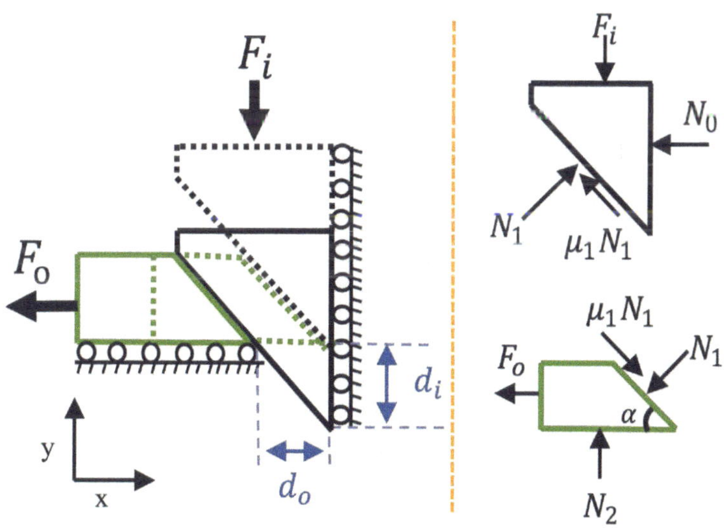

Figure 6.3 Illustration of static model for the wedge mechanism.

By combining Eqs. (6.1) through (6.4), the amplification ratio n_{wedge} can be obtained as:

$$n_{\text{wedge}} = \frac{F_o}{F_i} = \frac{\sin\alpha - \mu_1\cos\alpha}{\cos\alpha + \mu_1\sin\alpha} \tag{6.5}$$

When there is no friction force, that is, $\mu_1 = 0$, the equation can be derived:

$$n^0_{\text{wedge}} = \frac{F_o}{F_i} = \frac{d_i}{d_o} = \tan\alpha \tag{6.6}$$

where d_i and d_o are the input and output displacements, respectively. Force and displacement only depend on the tilt angle if there is no friction between the wedge structures.

6.3.2 Model of inner force amplifier

The analytical model of the inner force amplifier is established based on the stiffness matrix method. Such an approach can accurately predict the output performance of a complex mechanism based on the elastic deformation of compliant flexure beams (Zhu et al., 2016, 2018). A schematic of the inner force amplifier is depicted in Fig. 6.4. By referring to a frame element with

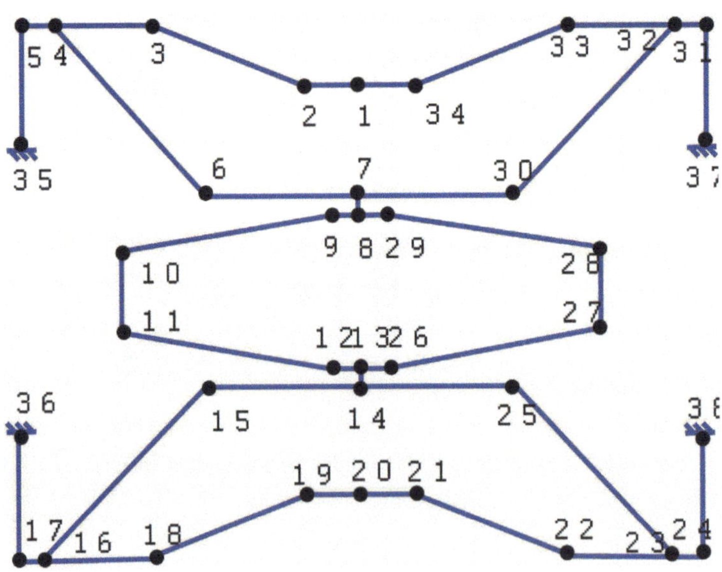

Figure 6.4 Schematic representation of inner force amplifier.

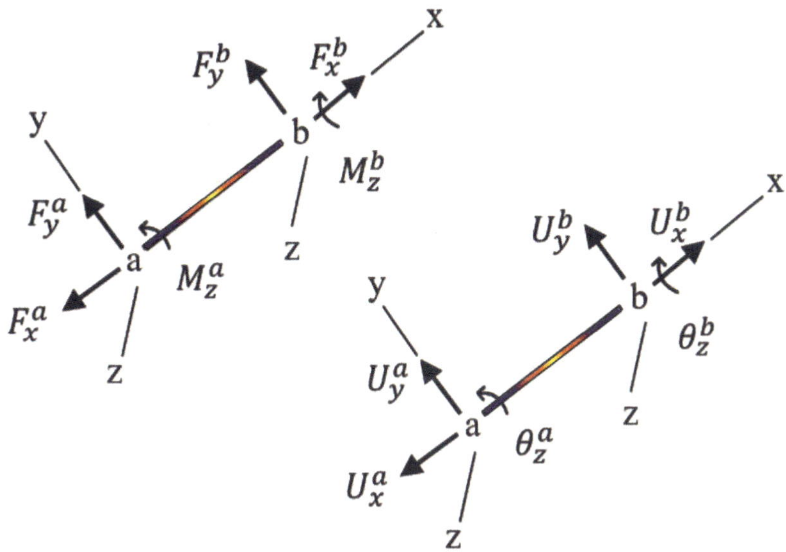

Figure 6.5 Illustration of frame element with two related nodes.

two related nodes, *a* and *b* (Fig. 6.5), the external force and displacement for each node can be expressed as:

$$\{F_a\} = \left[F_x^a\ F_y^a\ M_z^a\right]^T \tag{6.7}$$

$$\{F_b\} = \left[F_x^b F_y^b M_z^b\right]^T \tag{6.8}$$

$$\{U_a\} = \left[U_x^a\ U_y^a\ \theta_z^a\right]^T \tag{6.9}$$

$$\{U_b\} = \left[U_x^b\ U_y^b\ \theta_z^b\right]^T \tag{6.10}$$

where F_x^a, F_y^a, M_z^a, F_x^b, F_y^b, and M_z^b are the force components of nodes *a* and *b*; U_x^a, U_y^a, θ_z^a, U_x^b, U_y^b, and θ_z^b are the displacement components of nodes *a* and *b*; and superscript *T* represents the transpose of a vector or matrix.

The equation for the frame element can be written as:

$$\{F^e\} = \{K^e\}\{U^e\} \tag{6.11}$$

where $\{F^e\}$, $\{K^e\}$, and $\{U^e\}$ are the force, stiffness, and displacement components, respectively, of the frame element.

Substituting Eqs. (6.7)−(6.10) into Eq. (6.11), the following relationships can be further derived:

$$\{F_a\} = [K_{11}]\{U_a\} + [K_{12}]\{U_b\} \tag{6.12}$$

$$\{F_b\} = [K_{21}]\{U_a\} + [K_{22}]\{U_b\} \tag{6.13}$$

where $[K_{11}]$ and $[K_{22}]$ are symmetric matrices, and $[K_{12}] = [K_{21}]^T$ holds.

According to the relationship between the two nodes, local element stiffness matrices [K11] and [K22] can be obtained:

$$[K_{11}] = \begin{bmatrix} \frac{EA}{l} & 0 & 0 \\ 0 & \frac{12EI}{l^3} & \frac{6EI}{l^2} \\ 0 & \frac{6EI}{l^2} & \frac{4EI}{l} \end{bmatrix} \tag{6.14}$$

$$[K_{12}] = \begin{bmatrix} -\frac{EA}{l} & 0 & 0 \\ 0 & -\frac{12EI}{l^3} & \frac{6EI}{l^2} \\ 0 & -\frac{6EI}{l^2} & \frac{2EI}{l} \end{bmatrix} \tag{6.15}$$

where l is the length of the frame element, E is the Young's modulus, A is the sectional area, and I is the moment of inertia.

Moreover, to obtain the global element stiffness matrix, the rotational transformation matrix is considered, which can be written as:

$$[T] = \begin{bmatrix} \cos\beta & \sin\beta & 0 \\ -\sin\beta & \cos\beta & 0 \\ 0 & 0 & 1 \end{bmatrix} \tag{6.16}$$

Then, global stiffness matrix $\{K_T^e\}$ becomes the form:

$$\{K_T^e\} = [T]^T\{K^e\}[T] \tag{6.17}$$

Therefore, the equation for the inner amplification frame can be expressed as:

$$F = KU \tag{6.18}$$

where F and U represent the external force and displacement for all nodes, which can be further written into the forms:

$$U = \left[d^1 \ldots d^a \ldots d^n\right]^T \tag{6.19}$$

$$F = \left[F^1 \ldots F^a \ldots F^n\right]^T \tag{6.20}$$

where n denotes the number of nodes; and F^a and d^a are the force and displacement components of node a, respectively. The boundary condition is set by considering the situation of each node.

Finally, the amplification ratio for each submechanism can be determined according to the nodes, which is given as:

$$n_{\text{leverage}} = \frac{F^4}{F^1} \tag{6.21}$$

$$n_{\text{bridge}} = \frac{F^{10}}{F^8} \tag{6.22}$$

$$n_{\text{frame}} = \frac{F^{10}}{F^1} \tag{6.23}$$

where n_{leverage}, n_{bridge}, and n_{frame} are the amplification ratios for the leverage, bridge-type amplifier, and the whole inner amplification frame, respectively.

6.3.3 Total force amplification ratio

The total force amplification ratio for the energy harvester, $n_{\text{harvester}}$, becomes:

$$n_{\text{harvester}} = n_{\text{frame}} n_{\text{wedge}} \tag{6.24}$$

For the power output, the equivalent circuit of the piezoelectric stack is given in Fig. 6.6. The generated voltage under an open circuit and low frequency condition can be obtained according to the piezoelectric constitutive equation:

$$V_{\text{PZTstack}} = \frac{n_{\text{harvester}} F^1 g_{33}}{d} \tag{6.25}$$

where g_{33}, d, and V_{PZTstack} denote the voltage coefficient, width, and produced voltage of the piezoelectric stack, respectively.

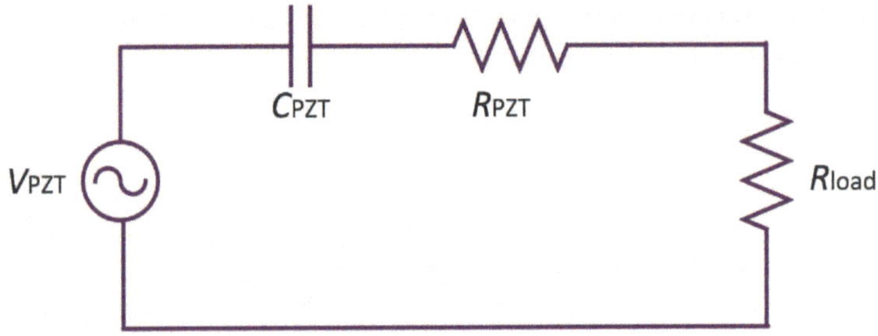

Figure 6.6 Equivalent circuit of a piezoelectric stack.

In addition, the created power $P_{\text{harvester}}$ can be calculated by the equation:

$$P_{\text{harvester}} = \frac{V_{\text{PZTstack}}^2}{R_{\text{load}}} \tag{6.26}$$

where R_{load} is the external loading resistance.

6.4 Parameter optimization and finite element analysis simulation study

Based on the derived analytical models, an optimal design of the structural parameters is carried out in this section. The performance of the optimized harvester is verified by conducting a finite element analysis (FEA) simulation study.

6.4.1 Dimension optimization

Because of the requirement of a potential practical application, such as capturing vibration energy from human footsteps, the input stroke is assigned to be 3 mm. Such a small magnitude is considered to ensure walking comfort and maintain high safety for users. For the wedge mechanism, the tilt angle (α) and friction coefficient (μ_1) are the key parameters; they have a significant impact on output performance. It is necessary to reduce the friction coefficient by applying appropriate lubricants. Otherwise, a high friction coefficient will restrict relative motion between the wedge components. Hence, its maximum value of 0.1 is considered in the design process.

According to Eq. (6.5), output forces under various tilt angles are generated and listed in Table 6.3. The wedge structure becomes a displacement amplifier when the tilt angle is less than 45 degrees. By

Table 6.3 Force output under various tilt angles of wedge structure.

α	15 degrees	30 degrees	45 degrees	60 degrees	75 degrees
μ_1	0.1	0.1	0.1	0.1	0.1
d_i (mm)	3	3	3	3	3
F_i (N)	100	100	100	100	100
d_0 (mm)	11.20	5.20	3.00	1.73	0.80
F_0 (N) with μ_1	16.35	45.13	81.82	139.11	264.49
F_0 (N) without μ_1	26.79	57.73	100.00	173.21	373.21

contrast, the wedge structure is able to amplify the input force when the tilt angle is greater than 45 degrees. Considering the improvement in the overall output performance and reliability of operation, the tilt angle is selected to be 60 degrees.

Moreover, multiparameter analysis of the inner force amplifier is carried out. It provides a comprehensive fundamental for the parameter optimization of the mechanism. Concerning the mechanical properties of each element, the flexure beams experience most stress under the action of input force. Therefore, the thickness (t), length (L), width (w), and height (h) of flexure beams should be considered.

In addition, regarding the leverage mechanism, the simplified input arm (L_1) refers to the distance between nodes 3 and 4. The simplified output arm (L_2) refers to the distance between nodes 4 and 5, as shown in Fig. 6.4. They are important parameters and affect the force amplification ratio of the leverage mechanism. Furthermore, to simplify the optimization process, one-quarter model of the inner force amplifier mechanism is adopted owing to its double symmetric structure. The height of the inner force amplifier is assigned to be 10 mm, to guarantee a compact footprint size and enough space for the motion of the wedge structures and screw fixture.

In this way, a total of 13 design parameters of the inner force amplifier mechanism are determined. The objective of the architectural optimization is to reduce the physical size and enlarge the force amplification ratio under the constraint of a high safety factor for the harvester. The details of the optimization progress are:

- Objective: To maximize the output force and force amplification ratio;
- Parameters: b_l^1, b_w^1, b_t^1, b_l^2, b_w^2, b_t^2, b_l^3, b_t^3, b_l^4, b_w^4, b_t^4, L_1, and L_2;
- Subject to:

  (1) Input displacement = 3 mm;

(2) Safety factor > 1.6;
(3) L_1, $L_2 \in$ [3:0.1:12] mm;
(4) b_l^1, b_l^2, $b_w^2 \in$ [5:0.1:12] mm;
(5) b_w^1, $b_l^3 \in$ [4:0.1:6] mm;
(6) b_t^1, b_t^2, b_t^3, $b_t^4 \in$ [0.5:0.1:1.5] mm;
(7) $b_l^4 \in$ [14:0.1:16] mm;
(8) $b_w^4 \in$ [2.5:0.1:3.5] mm.

Concerning the optimization algorithm, the method of MOGA can provide a global optimum with an acceptable accuracy for the multiobjective and multiconstraint optimization problem. Hence, the MOGA is adopted to optimize design parameters by using a built-in optimization toolbox in ANSYS software. The settings based on the toolbox are default values. Specifically, optimization begins with 800 samples and generates 80 samples per iteration. In addition, the one-point discrete crossover is used for the algorithm. The initial values and obtained optimal results are listed in Table 6.4.

6.4.2 Finite element analysis simulation results

To verify the optimal design of the harvester, an FEA simulation study is performed to evaluate the output performance of the energy harvester using ANSYS software under a frictionless situation. In addition, the piezoelectric stack and IMS frame are assigned as hexahedron mesh (with an element size of 0.5 mm) and tetrahedron mesh (with an element size of 0.5 mm), respectively.

Table 6.4 Optimization results of harvester parameters.

Parameter	Symbol	Initial value	Optimal value
Length of b^1 (mm)	b_l^1	10	11.4
Width of b^1 (mm)	b_w^1	5	5
Thickness of b^1 (mm)	b_t^1	0.8	0.5
Length of b^2 (mm)	b_l^2	10	9.8
Width of b^2 (mm)	b_w^2	10	10
Thickness of b^2 (mm)	b_t^2	0.8	0.6
Length of b^3 (mm)	b_l^3	5	6
Thickness of b^3 (mm)	b_t^3	0.8	0.5
Length of b^4 (mm)	b_l^4	15	16
Width of b^4 (mm)	b_w^4	3	2.7
Thickness of b^4 (mm)	b_t^4	0.8	1.1
Input arm (mm)	L_1	10	10
Output arm (mm)	L_2	4	4.9

Specifically, a displacement input of 3 mm is exerted on the top plate and the bottom plate is fixed. The reaction force is obtained based on the input ends of the piezoelectric stack. The simulation result of stress distribution of the optimized energy harvester is shown in Fig. 6.7. The output performance is listed in Table 6.5.

The wedge structure is able to bear much stress owing to its specific characteristics. The maximum stress occurs on the bridge-type force

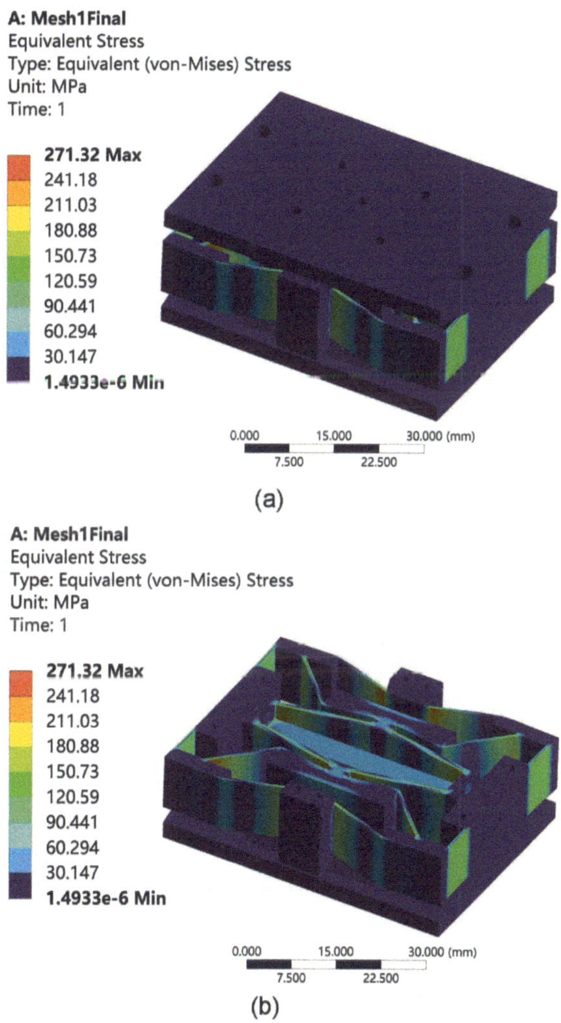

Figure 6.7 Simulation results of stress distribution for optimally designed energy harvester. (A) The entire harvester; (B) the harvester without a top plate.

Table 6.5 Performance of the designed integrated multistage harvester compared with other piezoelectric stack-based energy harvesters.

Performance	This work	Single two-stage harvester (Wen et al., 2018)	Compound two-stage harvester (Wen et al., 2018)	Two-stage harvester (Wang et al., 2017a)
Safety factor	1.98	1.23	2.94	1.60
Maximum stress (MPa)	271.32	410.62	170.88	1000
Displacement input (mm)	3	4.21	1.31	–
Equivalent force input (N)	67.73	100	100	–
Reaction force (N)	1275	2262	1790	–
Force amplification ratio by finite element analysis	18.83	22.62	17.90	21
Force amplification ratio by experiment	17.90	–	17.50	21
Size: $L \times W \times H$ (mm^3)	$53 \times 39.2 \times 25$	$60 \times 44 \times 34$	$60 \times 44 \times 34$	–
With protection mechanism	Yes	No	No	No

amplifier of b^4 (Fig. 6.1C) because magnified force is applied to the flexure beam. Moreover, the equivalent input force corresponding to 3-mm input displacement is obtained as 67.73 N. Because the IMS energy harvester has the function of position protection, a safety factor of 1.98 is enough for practical application. In addition, the simulation results reveal that the maximum output force is 1275 N and the total force amplification ratio is 18.83.

Table 6.5 shows that compared with existing designs based on a piezoelectric stack, the energy harvester allows a more compact physical size along with a protection mechanism to bear a large external load.

6.5 Experimental results and discussion

For experimental verification of the presented design, a prototype of the optimally designed energy harvester is fabricated, and experimental studies are carried out in this section.

6.5.1 Prototype fabrication

The fabricated prototype of the IMS energy harvester and the experimental setup are shown in Fig. 6.8. The input stroke is measured by a displacement

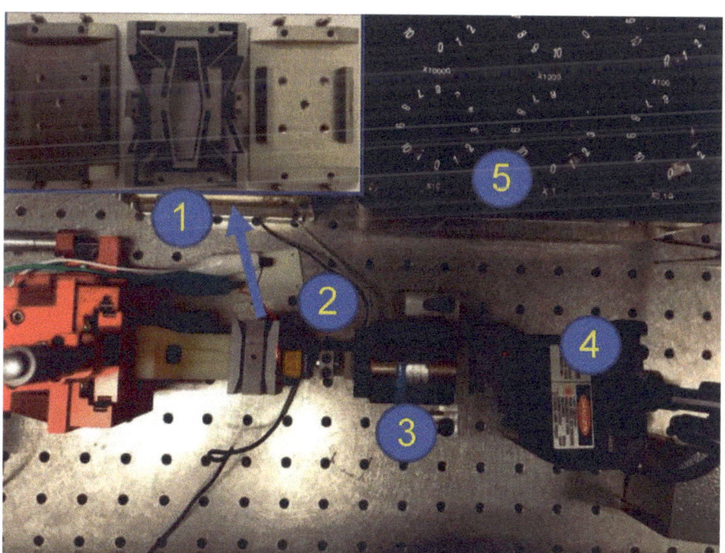

Figure 6.8 Fabricated prototype and experimental setup of energy harvester. (1) Prototype of fabricated integrated multistage energy harvester; (2) force sensor; (3) voice coil actuator; (4) laser displacement sensor; (5) variable resistance box.

sensor (LK-H055, Keyence Inc.) and analyzed using a multifunction data acquisition board (PXIe-6363, National Instruments Inc.). To evaluate the maximum power output, a variable resistance box (ZX21A, Shanghai Precision Instruments, Inc.) is applied to obtain optimal resistance for the IMS energy harvester. A force sensor (LSB200, FUTEK Advanced Sensor Technology, Inc.) and strain/bridge input module (PXIe-4330, National Instruments, Inc.) are used to acquire external force exerted on the harvester. In addition, a digital multimeter (PXIe-4082, National Instruments, Inc.) is introduced to measure voltage generated by the piezoelectric stack. A real-time control (PXIe-1082, National Instruments, Inc.) is adopted to collect data. Moreover, a voice coil actuator (NCC05-18-060-2X, H2W Techniques, Inc.) is used to provide periodic excitation motion to drive the IMS energy harvester.

6.5.2 Experimental results

First, the complete press-and-release cycle of the IMS energy harvester is tested by applying a compressive force of 82.29 N. This force is larger than the calculated equivalent input force (with 3-mm displacement input) to ensure that the device is fully compressed. The input signals are indicated in Fig. 6.9 and the output performance is shown in Fig. 6.10. A maximum input displacement of 2.89 mm is measured. Moreover, the external force acting on the IMS energy harvester can produce a large power output. In particular, a peak power of 34.81 mW and peak voltage of 26.39 V are generated across a 20-kΩ resistor. Furthermore, because the elastic beams restore the original shape when force is removed, a negative voltage is generated during this process.

Fig. 6.10A shows that the voltage output process of the mechanism can be classified into three stages.

The first stage refers to displacement between 0 and 3 mm. The voltage is positively proportional to the input force. The second stage corresponds to displacement at 3 mm. No further force is allowed when displacement reaches 3 mm. Because the input force remains constant, the piezoelectric stack does not generate more power. Hence, voltage dramatically decreases and returns to the zero in a short time. The third stage refers to displacement from 3 to 0 mm. Because the input force decreases, it leads to a negative value for voltage. The negative voltage is also positively proportional to the input force until the force is released. However, the restoring force is relatively smaller than the compressive force because a large amount

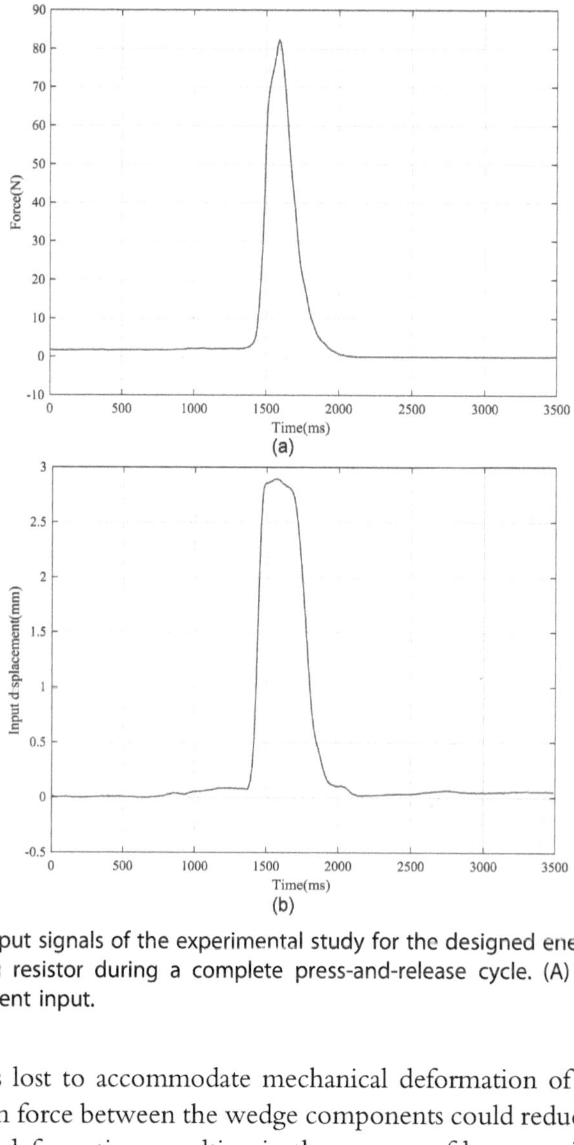

Figure 6.9 Input signals of the experimental study for the designed energy harvester with a 20-kΩ resistor during a complete press-and-release cycle. (A) Force signal; (B) displacement input.

of energy is lost to accommodate mechanical deformation of the device. Also, friction force between the wedge components could reduce the speed of recovery deformation, resulting in the capture of less restoring energy.

Second, the force amplification ratio of the IMS energy harvester is tested by measuring differences in the slope of fitting curves for the harvester and piezoelectric stack, as shown in Fig. 6.11. The generated voltage is proportional to the exerted force. The results indicate that the force amplification ratio of the device is 17.9. The nonlinear effect of the piezoelectric stack in the IMS is not considered here. In addition, based on

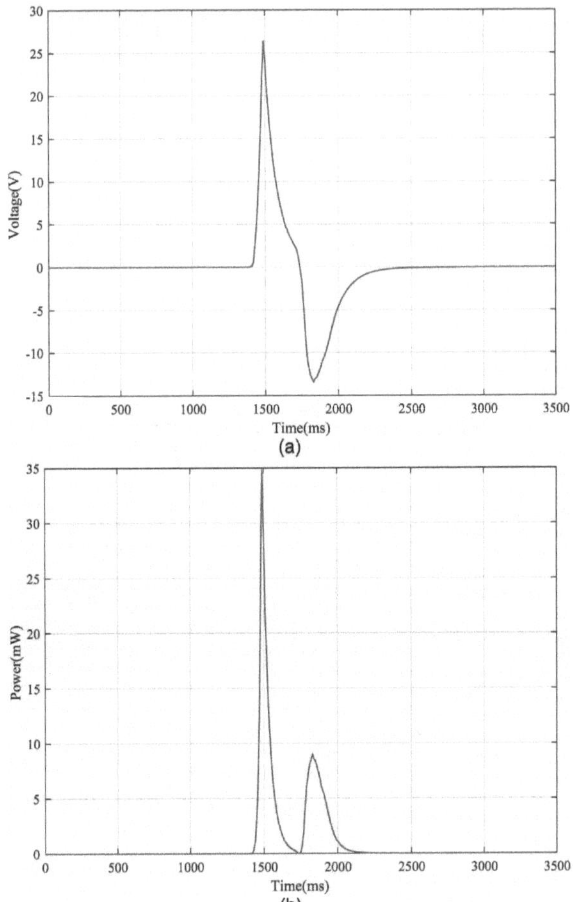

Figure 6.10 Experimental results of output performance for the designed energy harvester with a 20-kΩ resistor during a complete press-and-release cycle. (A) Generated voltage signal; (B) generated power.

the collected data, the curve fitting results are shown in Fig. 6.11. The nonzero value of DC component may be caused by errors in the measurement data. Moreover, the adopted measurement equipment may result in unexpected voltage that is generated. Because this value is small, its influence on the result is not considered here.

Third, the optimal output performance of the harvester under various resistances at an excitation frequency of 5 Hz is tested. The input signals are shown in Fig. 6.12 and the output power results are shown in Fig. 6.13. The average output power versus different resistance value is depicted in

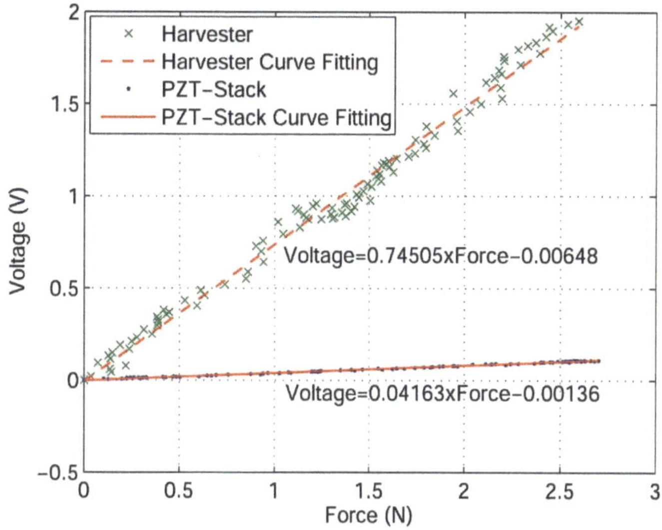

Figure 6.11 Relationships between force and voltage for the integrated multistage energy harvester and piezoelectric stack.

Fig. 6.13B. The optimal resistance Z_p of the piezoelectric stack (regarded as pure capacitance) can be approximated by (Liao & Sodano, 2008):

$$Z_p = \frac{1}{\omega_p C_p} \tag{6.27}$$

where ω_p and C_p are the excitation frequency and piezoelectric capacitance, respectively.

For the resistance matching theory, maximum power occurs when Z_p is equal to the matched resistance. Experimental results show that a maximum peak power of 50.8 mW and average power of 13.15 mW are obtained across a 10-kΩ resistor under an input force of 65.73 N. Optimal resistance is close to the calculated resistance value of 10.27 kΩ. Moreover, the maximum input displacement is measured at 2.76 mm. Because of the influence of friction force, the device cannot be fully compressed when the applied force is close to the equivalent input force. However, the friction force imposes no significant influence on the output performance.

In addition, to illustrate its practical utility, the IMS harvester is excited to light 50 LEDs by the footsteps of a person weighing 65 kg. Each LED has a diameter of 3 mm. A snapshot of the experimental results is given in Fig. 6.14. The result further demonstrates that the proposed IMS energy harvester is able to bear a large load and generate a large output energy.

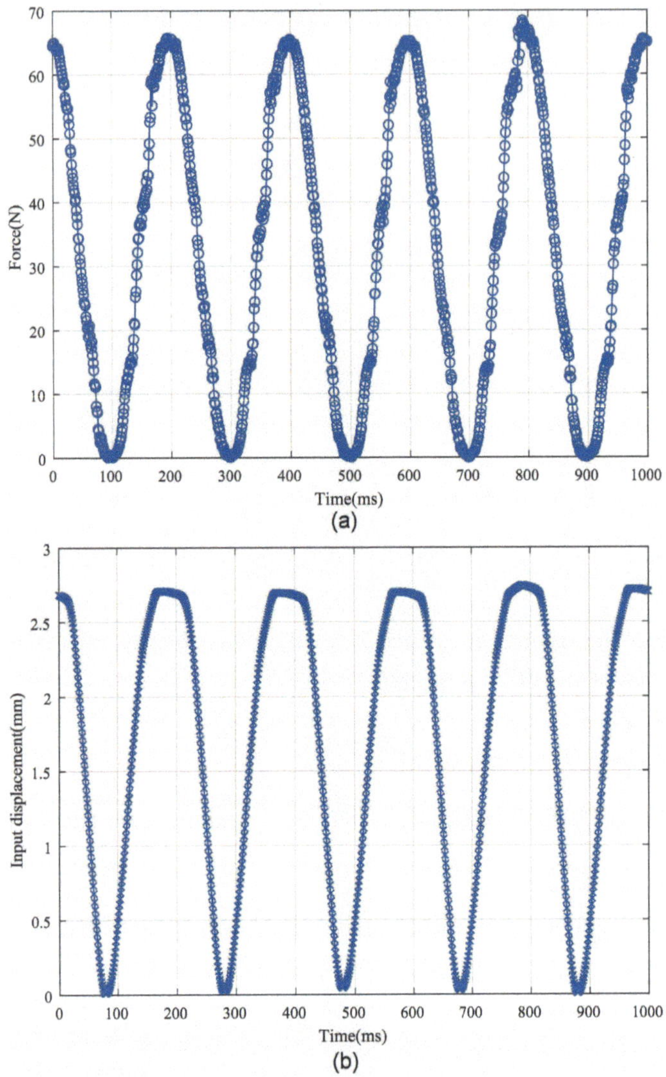

Figure 6.12 Input signal for harvester testing under various resistances. (A) Force signal; (B) displacement input.

6.5.3 Further discussion

An objective of the proposed mechanism design is to ensure a purely vertical motion of the harvester excited by human walking. It is best to obtain maximum output power in the working direction, whereas forces in other directions are restricted. Hence, the safety of the device material is enhanced.

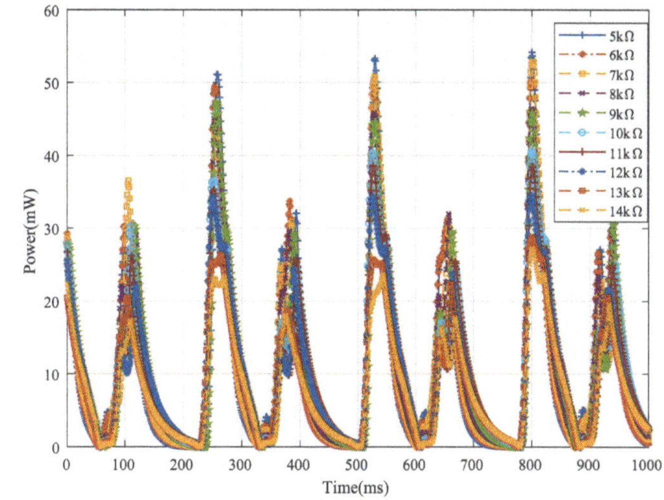

(a)

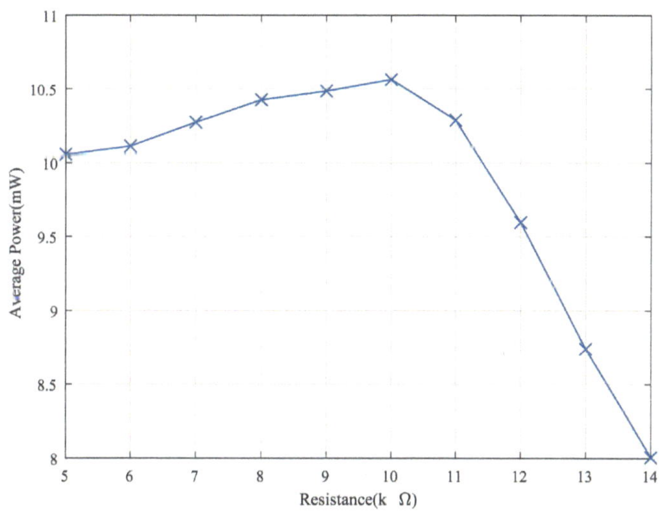

(b)

Figure 6.13 Output performance of harvester under various resistances. (A) Power output; (B) average power output versus resistance.

The experimental result of the device demonstrates how an input displacement of 3 mm leads to maximum output power. The input force is not directly concerned because it is the equivalent force induced by 3-mm input displacement. However, the input force is not expected to be too large because of the limitation of the frame's material limit. A large input

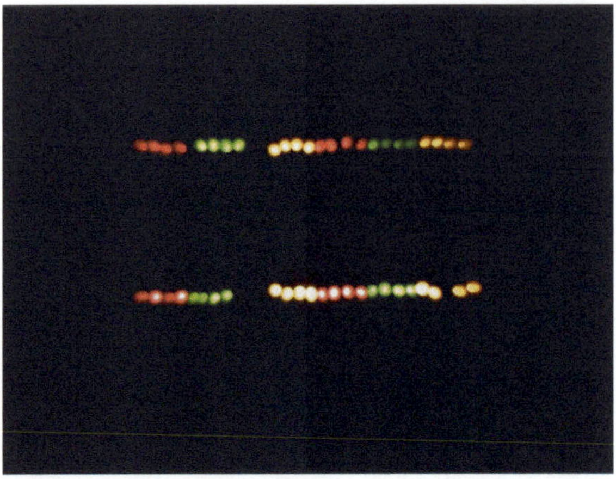

Figure 6.14 Experimental result of energy harvester lighting 50 LEDs 3 mm in diameter, which are excited by the footsteps of a person weighing 65 kg.

force could lead to a longer and thicker dimension of the elastic beam, resulting in a larger physical size of the frame. In addition, the output force is restricted. Thus, a smaller input force is more suitable for a design that could improve the simplification and feasibility of the device.

To obtain the maximum output power, a large output force is required. Because the force amplification ratio relates the output force to the input force, when the input force is constrained, a larger ratio is desired. Hence, a large amplification ratio is essential. Therefore, the target is to maintain a large output power when the input displacement is smaller than 3 mm. There is certain critical point on the output force that is exactly the critical value for the piezoelectric stack before damage. Thus, during the mechanism design of the energy harvester, it is essential to guarantee the harvester's safety, stability, and reliability, and then to guarantee maximum output power.

The drawback of this mechanism is that flexure beams could bear only limited stress. This results in a decrease in the safety factor and restricts the force amplification ratio accordingly. Otherwise, the force amplification ratio could be large based on the multistage force amplification mechanism. As an effective approach to enhancing the safety factor, the single flexure beam could be replaced with a compound flexure beam. However, it also increases the total physical size of the harvester. Moreover, the equivalent input force would be increased owing to an increase in the input stiffness of the device. Hence, the structure optimization of the energy harvester should be conducted considering the requirement of different practical situations.

In practice, the mass of the prototype harvester is about 73 g. The proposed energy harvester can be buried underground to scavenge energy from human footsteps during walking. Most important, owing to the position protection mechanism, external force does not affect the structure's safety. Thus, the experience of walking can be satisfied according to constraints on displacement input. Although the required input force (around 80 N) seems small, the maximum force exerted on both sides of the piezoelectric stack (1432 N) is 17.9 times the input force. It is enough to scavenge a large amount of energy. With the imposition of a heavy person (65 kg), the harvester is not damaged owing to the proposed protection mechanism design. However, the weight of the person is not fully converted into electrical power by the single harvester. In practice, owing to the compact size of the design, an array of the harvesters can be mounted under the floor to scavenge mechanical energy more efficiently from human footsteps. In this way, the person's weight is divided and supported by multiple harvesters. The person's weight is fully used by each harvester to produce electrical power.

Because this work mainly focuses on the mechanical design of the energy harvester, electrical parts are of less concern. The experimental study was carried out to test the output performance of the harvester. LEDs were tested in the experiment for demonstration. The objectives were to verify that the device could withstand a force beyond the harvester's limit and to validate that the output power was high enough to light 50 LEDs. As future work, a more sophisticated electrical circuit design will be studied to improve the harvesting ability.

6.6 Conclusion

This chapter presents the design, modeling, optimization, and testing of a piezoelectric energy harvester based on an integrated multistage force amplification mechanism. The designed energy harvester integrates four types of force amplification frames: the wedge mechanism, leverage mechanism, half-bridge mechanism, and bridge-type force amplifier. The analytical models are established based on force analysis and the stiffness matrix method. The main design variables of the models are optimized to achieve the best force amplification ratio under a large safety factor. Using the force amplifier, the energy harvester demonstrates higher output performance, a larger safety factor, and a more compact size than previous work on piezostack-based energy harvesters. Experimental investigations demonstrate the effectiveness and utility of the developed energy harvester.

References

Feenstra, J., Granstrom, J., & Sodano, H. (2008). Energy harvesting through a backpack employing a mechanically amplified piezoelectric stack. *Mechanical Systems and Signal Processing, 22*(3), 721–734. https://doi.org/10.1016/j.ymssp.2007.09.015

Liao, Y., & Sodano, H. A. (2008). Model of a single mode energy harvester and properties for optimal power generation. *Smart Materials and Structures, 17*(6). https://doi.org/10.1088/0964-1726/17/6/065026

Liu, P., & Yan, P. (2016). A new model analysis approach for bridge-type amplifiers supporting nano-stage design. *Mechanism and Machine Theory, 99*, 176–188. https://doi.org/10.1016/j.mechmachtheory.2016.01.005

Su, S. X. P., & Yang, H. S. (2002). Analytical modeling and FEM simulations of single-stage microleverage mechanism. *International Journal of Mechanical Sciences, 44*(11), 2217–2238. https://doi.org/10.1016/S0020-7403(02)00162-5

Wang, L., Chen, S., Zhou, W., Xu, T. B., & Zuo, L. (2017a). Piezoelectric vibration energy harvester with two-stage force amplification. *Journal of Intelligent Material Systems and Structures, 28*(9), 1175–1187. https://doi.org/10.1177/1045389X16667551

Wang, Y., Ding, H., Le, X., Wang, W., & Xie, J. (2017b). A MEMS piezoelectric in-plane resonant accelerometer based on aluminum nitride with two-stage microleverage mechanism. *Sensors and Actuators, A: Physical, 254*, 126–133. https://doi.org/10.1016/j.sna.2016.12.019

Wang, X., Shi, Z., Wang, J., & Xiang, H. (2016). A stack-based flex-compressive piezoelectric energy harvesting cell for large quasi-static loads. *Smart Materials and Structures, 25*(5). https://doi.org/10.1088/0964-1726/25/5/055005

Wen, S., & Xu, Q. (2018). Design of a two-stage force amplification frame for piezoelectric energy harvesting. In *2017 IEEE international conference on cybernetics and intelligent systems, CIS 2017 and IEEE conference on robotics, automation and mechatronics, RAM 2017 - proceedings* (Vol. 2018, pp. 490–495). Institute of Electrical and Electronics Engineers Inc. https://doi.org/10.1109/ICCIS.2017.8274825

Wen, S., Xu, Q., & Zi, B. (2018). Design of a new piezoelectric energy harvester based on compound two-stage force amplification frame. *IEEE Sensors Journal, 18*(10), 3989–4000. https://doi.org/10.1109/JSEN.2018.2820221

Xie, L., & Cai, M. (2014). Increased piezoelectric energy harvesting from human footstep motion by using an amplification mechanism. *Applied Physics Letters, 105*(14). https://doi.org/10.1063/1.4897624

Xu, T. B., Siochi, E. J., Kang, J. H., Zuo, L., Zhou, W., Tang, X., & Jiang, X. (2013). Energy harvesting using a PZT ceramic multilayer stack. *Smart Materials and Structures, 22*(6). https://doi.org/10.1088/0964-1726/22/6/065015

Yang, Z., & Zu, J. (2016). Toward harvesting vibration energy from multiple directions by a nonlinear compressive-mode piezoelectric transducer. *IEEE, 21*(3), 1787–1791. https://doi.org/10.1109/TMECH.2015.2459014

Zhou, W., & Zuo, L. (2013). A novel piezoelectric multilayer stack energy harvester with force amplification. In *Proceedings of the ASME 2013 international design technical conferences and computers and information in engineering conference*. ASME. V008T13A095.

Zhu, W. L., Zhu, Z., Guo, P., & Ju, B. F. (2018). A novel hybrid actuation mechanism based XY nanopositioning stage with totally decoupled kinematics. *Mechanical Systems and Signal Processing, 99*, 747–759. https://doi.org/10.1016/j.ymssp.2017.07.010

Zhu, W. L., Zhu, Z., Shi, Y., Wang, X., Guan, K., & Ju, B. F. (2016). Design, modeling, analysis and testing of a novel piezo-actuated XY compliant mechanism for large workspace nano-positioning. *Smart Materials and Structures, 25*(11). https://doi.org/10.1088/0964-1726/25/11/115033

CHAPTER 7

Design of a bidirectional energy harvester with a single piezoelectric stack

This chapter presents the design, fabrication, and testing of a novel piezoelectric energy harvester with a single piezoelectric stack to harvest energy from excitations with a bidirectional motion. A rhombus-type force amplifier is combined with a selectivity lever to use both pull and push inputs. An analytical model is established to predict the force amplification ratio of the proposed energy harvester under different directions of input force. The mechanism parameters are optimally designed to maximize the output performance of the energy harvester. A prototype energy harvester is fabricated for experimental testing. Results indicate that the developed harvester has a high safety factor and compact size, which can be adopted to harvest mechanical energy for excitations with both pull and push inputs bidirectionally.

7.1 Introduction

Many piezoelectric energy harvesters with a piezoelectric stack have been reported in the literature. However, most can scavenge energy only from a unidirectional input force (Chen et al., 2017; Feenstra et al., 2008; Wang et al., 2016; Wen et al., 2018; Xu et al., 2013). Thus, energy produced in other directions of excitation cannot be harvested and will be wasted. An energy harvester dedicated to more directions of input force is more practical. Although an energy harvester can easily be designed to scavenge energy from multiple directions of excitation using multiple piezoelectric stacks, an energy harvester with fewer piezoelectric stack is more economical in practice. From this point of view, it is desirable to develop a piezoelectric energy harvester with a single piezoelectric stack to scavenge energy from a bidirectional input force. Previously, a piezoelectric cantilever coupled with two permanent magnets was introduced to harvest

Mechanical Design of Piezoelectric Energy Harvesters
ISBN 978-0-12-823364-1
https://doi.org/10.1016/B978-0-12-823364-1.00004-3
© 2022 Elsevier Inc.
All rights reserved.

energy from bidirectional excitation with both attractive and repulsive forces (Challa et al., 2008). More recently, a bidirectional nonlinear piezoelectric energy harvester was developed (Fan et al., 2014), which was realized by two magnetically coupled piezoelectric cantilever beams. However, few bidirectional energy harvesters based on a piezoelectric stack were reported in the literature.

This chapter presents the design and development of a bidirectional piezoelectric energy harvester with a single piezoelectric stack to scavenge energy from both pull and push excitations. In particular, a compact piezoelectric energy harvester was devised. To apply the piezoelectric energy harvester with both pull and push inputs, a selectivity lever is introduced, composed of a lever mechanism and position limiters. The lever mechanism is adapted to change the direction of input force, and position limiters are employed to distinguish between pull and push inputs. When there is a pull input, the input force will be directly transmitted to the force amplifier. Otherwise, when there is a push input, the input force will first be transmitted to the lever mechanism and the force amplifier will obtain the input force after the change in direction. To generate a higher output power, a rhombus-type amplifier is adopted to amplify the input force. To maximize the output performance of the energy harvester, a multiobjective genetic algorithm (MOGA) method is used to optimize the parameters of the flexure mechanism based on finite element analysis (FEA). A prototype harvester was fabricated for experimental verification of the proposed conceptual design. The energy harvester can be mounted on the floor to scavenge energy excited by human walking, in which the bidirectional friction force generated between footsteps and the harvester is converted into electricity.

7.2 Mechanism design

In this section, the mechanism design of the proposed energy harvester is introduced in detail. The computer-aided design (CAD) model of the designed energy harvester is shown in Fig. 7.1, and its parameters are illustrated in Fig. 7.2. To facilitate the analysis, assign F_{input} ($f_{y,a}$) as the input force, F_{PZT} ($f_{x,f}$) as the output force, Δy_a as the input displacement, and Δx_f as the output displacement of the mechanism.

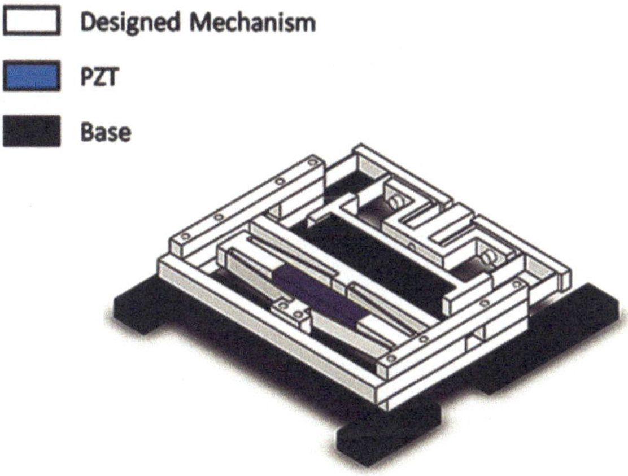

Figure 7.1 Computer-aided design model of proposed energy harvester.

7.2.1 Working principle

As shown in Fig. 7.2, when a pull input is applied to part *a* (Fig. 7.3), position limiter A will constrain the deflection of the flexure hinges of the selectivity lever to avoid material failure. At the same time, input force is transmitted to the rhombus-type amplifier through the lever mechanism. On the contrary, when there is a push input applied to part *a* (Fig. 7.4), position limiter B will act as a stopper. Meanwhile, the lever mechanism will change the direction of input force at first and the input force is

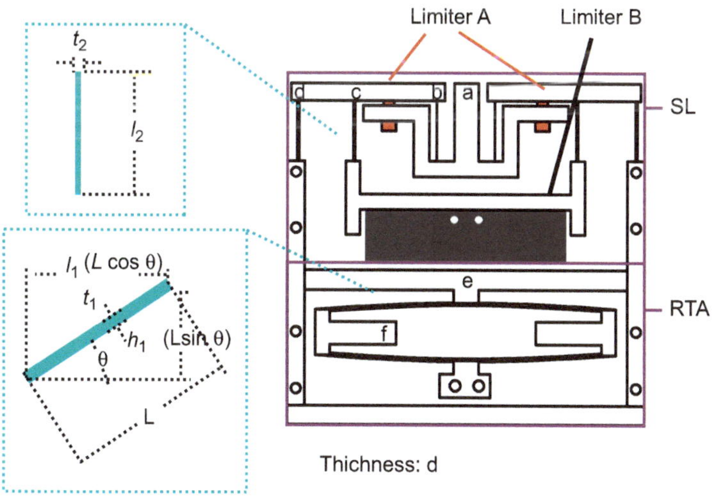

Figure 7.2 Illustration of parameters for the designed energy harvester.

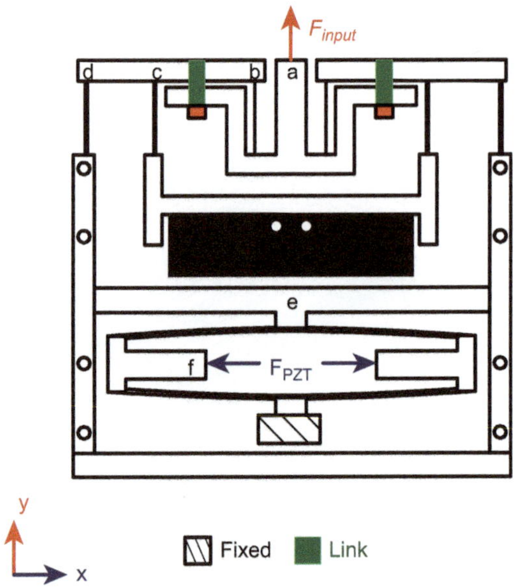

Figure 7.3 Schematic diagram of mechanism design under pull input.

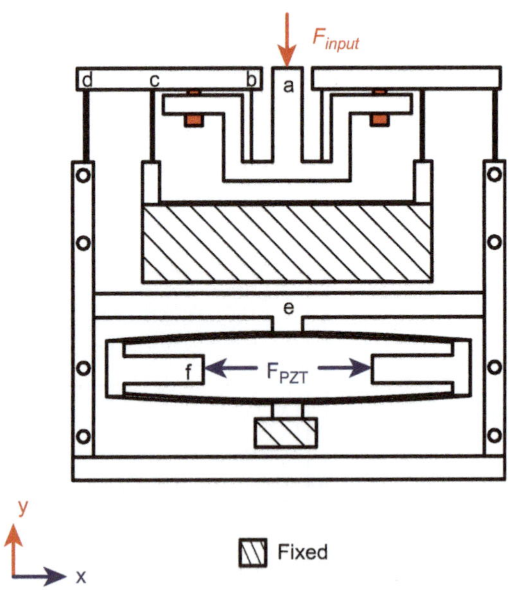

Figure 7.4 Schematic diagram of mechanism design under push input.

transmitted to rhombus-type amplifier after the change in direction. In this way, both pull and push force inputs will be transmitted to the piezoelectric stack located inside the rhombus-type amplifier. The force amplifier contributes to improving the output power of the piezoelectric energy harvester.

7.2.2 Model of mechanism under pull input

When there is a pull input, position limiter A takes effect; it can be treated as a link between components *a* and *b*. Therefore, the mechanism design under a pull input is simplified, as shown in Fig. 7.3. The relationships of the mechanism under a pull input can be determined by:

$$y_a = y_e \tag{7.1}$$

$$F_{\text{input}} = f_{y,e} \tag{7.2}$$

where y_a and y_e are the y-axis displacements of parts *a* and *e*, respectively. $f_{y,e}$ is the y-axis force of part *e*.

As shown in Fig. 7.3, the mechanism can be treated as a rhombus-type amplifier. Because the rhombus-type amplifier is symmetric, it needs to model only a quarter of the structure. The static analysis is illustrated in Fig. 7.5A, which gives:

$$2M_{ef} = f_{x,f} \times L \sin\theta - f_{y,e} \times L \cos\theta \tag{7.3}$$

where M_{ef} is the moment supplied by flexure hinge *ef* to keep a static balance state. L and θ are the structural parameters of the rhombus-type amplifier, as shown in Fig. 7.2.

For a rhombus-type amplifier, the deflection component in the y-direction is much smaller than that in the x-direction. Thus, deflection in the neutral axis is treated as x. Corresponding moment M_{ef} and axial tension f_{ef} in the elastic beam of flexure hinge *ef* can be calculated as:

$$M_{ef}(x) = M_{ef} - \left(f_{x,f}\sin\theta - f_{y,e}\cos\theta\right)x \tag{7.4}$$

$$f_{ef}(x) = f_{x,f}\cos\theta + f_{y,e}\sin\theta \tag{7.5}$$

Assume that flexure hinge *ef* is a Euler-Bernoulli beam and deflection is caused by input force $f_{y,e}$. Based on the principle of the conservation of energy, the energy equation of flexure hinge *ef* is derived as:

$$\frac{1}{2}f_{x,f}\Delta x_{ef} - \frac{1}{2}f_{y,e}\Delta y_{ef} = \int_0^L \frac{f_{ef}(x)^2}{2EA_{ef}}dx + \int_0^{L_1} \frac{M_{ef}(x)^2}{2EI_{ef}}dx \tag{7.6}$$

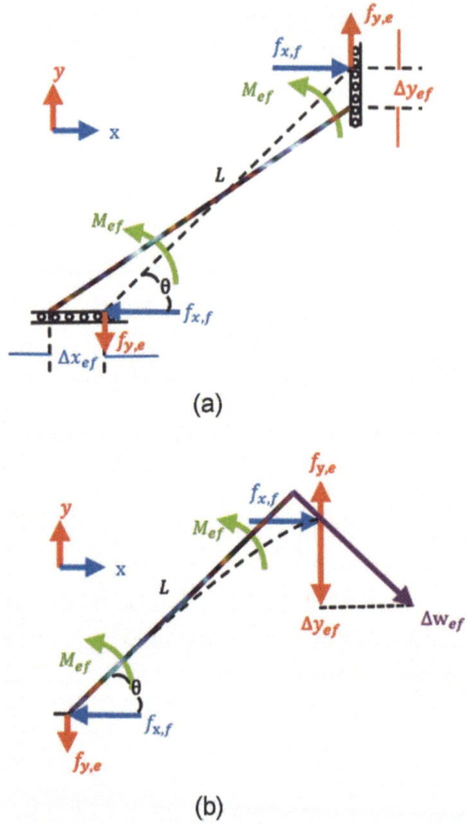

Figure 7.5 (A) Schematic diagram and (B) deflection of one-quarter of rhombus-type amplifier.

where I_{ef} and A_{ef} are the moment of inertia and area of the cross-section of flexure hinge *ef*, respectively. E is the Young's modulus of the material.

According to the Euler-Bernoulli beam theory, transverse displacement Δw, where d and t_1 are the structural parameters of the rhombus-type amplifier, as shown in Fig. 7.2, of the neutral axis, can be derived by:

$$\Delta w_{ef} = \int_0^L \int_0^x \frac{M_{ef}(u)}{2EI_{ef}} dudx = \frac{\left(f_{x,f} \sin\theta - f_{y,e} \cos\theta\right) L_1^3}{Edt_1^3} \tag{7.7}$$

As shown in Fig. 7.5B, output displacement Δy_{ef} is along the direction of $f_{y,e}$, which is the vertical component of Δw_{ef}. The relationship is described as:

$$\Delta y_{ef} = \Delta w_{ef} \cos\theta = \frac{\left(f_{x,f} \sin\theta - f_{y,e} \cos\theta\right) L^3 \cos\theta_1}{Edt_1^3} \tag{7.8}$$

In view of Eqs. (7.3)–(7.8), Δx can be derived as:

$$\Delta x_{ef} = \frac{\left(f_{x,f} \sin\theta - f_{y,e} \cos\theta\right) L^3 \sin\theta}{Edt_1^3} + \frac{\left(f_{x,f} \cos\theta + f_{y,e} \sin\theta\right)^2 L}{f_{x,f} Edt_1} \tag{7.9}$$

Let K_{PZT} represent the stiffness of the piezostack. Deflection of the piezostack can be determined as $2\Delta x_{ef} = f_{x,f}/K_{\text{PZT}}$. Then, submitting it into Eq. (7.9), the force amplification ratio under a pull input can be derived as:

$$R_{\text{pull}} = \frac{f_{x,f}}{F_{\text{input}}} = \frac{-b + \sqrt{b^2 - 4ac}}{2a} \tag{7.10}$$

where $a = \cos^2\theta + \frac{L^2}{t_1^2}\sin^2\theta - \frac{1}{2K_{\text{PZT}}}$, $b = \sin\theta\cos\theta\left(2 - \frac{L^2}{t_1^2}\right)$, and $c = \sin^2\theta$.s.

7.2.3 Model of mechanism under push input

When there is a push input, position limiter B takes effect; it can be treated as the fixed part of the lever mechanism. Therefore, the mechanism design under a push input is simplified as shown in Fig. 7.4. Because the model of hinge *ef* is the same as before, the model of hinge *ef* can be generated as:

$$R_{ef} = R_{\text{pull}}, \; y_e = 2\Delta y_{ef} \tag{7.11}$$

where R_{ef} is the force amplification ratio of hinge *ef*.

Furthermore, another part of the mechanism, which is a kind of lever mechanism, is shown in Fig. 7.6. Assume that hinges *de*, *ab*, and *c*-fixed have the same shape. These can be treated as a parallel four-bar mechanism, and the relationships are governed by:

$$f_{y,d} = f_{y,e}, \; f_{y,b} = F_{\text{input}}, \; \Delta y_{cd} = y_e, \; \theta_{bd} = \arcsin\left(\frac{\Delta y_{cd}}{l_{cd}}\right) \tag{7.12}$$

where $f_{y,d}$ and $f_{y,b}$ are the y-axis forces of parts *d* and *b*, respectively. Δy_{cd} is the y-axis displacement between parts *c* and *d*. θ_{bd} is the angle between beam *bd* and the x-axis. l_{cd} is the displacement between parts *c* and *d*.

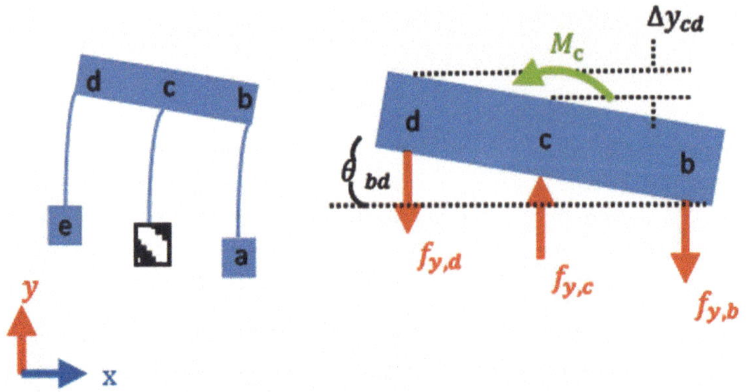

Figure 7.6 Schematic diagram of lever mechanism.

Based on the Euler-Bernoulli beams of hinges *de*, *ab*, and *c*-fixed, moment M_c can be determined as (Lobontiu et al., 2001):

$$M_c = \frac{Edt_2^3\theta_{bd}}{12l_2} \tag{7.13}$$

where l_2 and t_2 are the structural parameters of the selectivity lever, as shown in Fig. 7.2.

According to the moment balance principle in the static situation, the governing equation for the lever mechanism can be determined as:

$$M_c + f_{y,d}l_{cd}\cos\theta_{bd} = f_{y,b}l_{bc}\cos\theta_{bd} \tag{7.14}$$

where l_{bc} is the displacement between the two parts, *b* and *c*.

Considering Eq. (7.12)−(7.14), force $f_{y,e}$ can be calculated as:

$$f_{y,e} = \frac{F_{\text{input}}l_{bc}}{l_{cd}} - \frac{Edt_2^3\theta_{bd}}{12l_2l_{cd}\cos\theta_{bd}} \tag{7.15}$$

Then, the force amplification ratio under a push input force can be determined as:

$$R_{\text{push}} = R_{ef}\cdot\frac{f_{y,e}}{F_{\text{input}}} \tag{7.16}$$

7.2.4 Model of piezoelectric stack

Assume that the acting force of the piezostack is harmonic force $F(t) = F_0e^{-j\omega t}$, where F_0 is the amplitude of the acting force. Complex voltage V can be determined as (Zhao & Erturk, 2014):

$$V = \frac{j\omega d_{33}^{eff}F_0}{j\omega C_p^{eq} + (1/R_l)} \tag{7.17}$$

where d_{33}^{eff}, C_p^{eq}, and R_l are the effective piezoelectric constant, equivalent capacitance, and resistance of load, respectively.

For Eq. (7.17), open-circuit voltage V_{open} and output power P can be derived as:

$$V_{open} = \frac{d_{33}^{eff} F_0}{C_p^{eq}} \tag{7.18}$$

$$P = \frac{|V|^2}{R_l} = \frac{\left(\omega d_{33}^{eff} F_0\right)^2}{\left(\omega C_p^{eq}\right)^2 R_l + 1/R_l} \tag{7.19}$$

According to Eq. (7.19), the output power can be maximized when the load resistance is $R_l = 1\Big/\left(\omega C_p^{eq}\right)$. Thus, the maximum output power can be calculated as:

$$P_{max} = \frac{\omega\left(d_{33}^{eff} F_0\right)^2}{2C_p^{eq}} \tag{7.20}$$

7.3 Parameter optimization

In this section, the parameters of the proposed harvester mechanism are optimized by the MOGA method based on FEA simulation. The selection of variable parameters, constraints of FEA, and optimization results are presented next.

7.3.1 Variable parameter selection

To implement optimization, the parameters that affect the performance of the proposed energy harvester should be selected and optimized. Considering the size of the piezostack, the thickness of structure d is assigned to be 5 mm. Then, referring to Fig. 7.2, the variable parameters are determined to be:

$$\text{Parameter}(l_1, h_1, t_1, l_2, t_2, l_{cd}, l_{bd}) \tag{7.21}$$

where parameters l_1, h_1, t_1, l_2, and t_2 are denoted in Fig. 7.2. In addition, l_{cd} is the displacement between parts c and d, and l_{bd} represents the displacement between parts b and d.

7.3.2 Optimization setup and result

In this work, the MOGA method based on FEA simulation (with ANSYS Workbench software) is adopted for optimization. Before optimization is implemented, the model of the mechanism is parameterized to make the structure change based on the defined parameter relationships. In addition, the material of the mechanism is selected to be Al-7075 because of its high elasticity.

Two static structural analyses are carried out with the pull and push input forces. A large input force will cause a large physical size of the proposed harvester. Thus, to facilitate machining and testing, the magnitudes of the input forces are both selected to be 50 N in the current work. The analysis settings are shown in Figs. 7.7A and 7.8A. To obtain a higher output energy, the key objective of the optimization is to maximize the sum of output forces of the two static structural analyses. In addition, to avoid material failure, it is necessary to set a constraint for the safety factor greater than 1.0 for both static structural analyses. In this case study, the safety factor is designed to be greater than 2.0. Limited by the assigned safety factor of the material, if we increase the input force, the optimal force amplification ratio will decrease. On the contrary, if we decrease the input force, the optimal force amplification ratio will increase.

Based on these constraints, the optimization results are obtained as shown in Table 7.1. The output force is equal to 610 N under a pull input of 50 N. In addition, the output force is 657 N under a push input of 50 N. Thus, the force amplification ratio is 12.20 and 13.14 under the pull and push inputs, respectively. The deflections of the structure are shown in Figs. 7.7B and 7.8B, and the safety factor distributions of the mechanism are shown in Figs. 7.7C and 7.8C.

7.4 Prototype fabrication and experimental testing

In this section, a prototype harvester with optimal structure parameters was fabricated to verify the performance of the proposed energy harvester.

7.4.1 Prototype fabrication and experimental setup

Based on the CAD model, the selectivity lever and rhombus-type amplifier were manufactured from Al-7075 material with a tolerance of ±0.05 mm. Other components were fabricated to satisfy the functional demands of the proposed design. The prototype and experimental setup are shown in

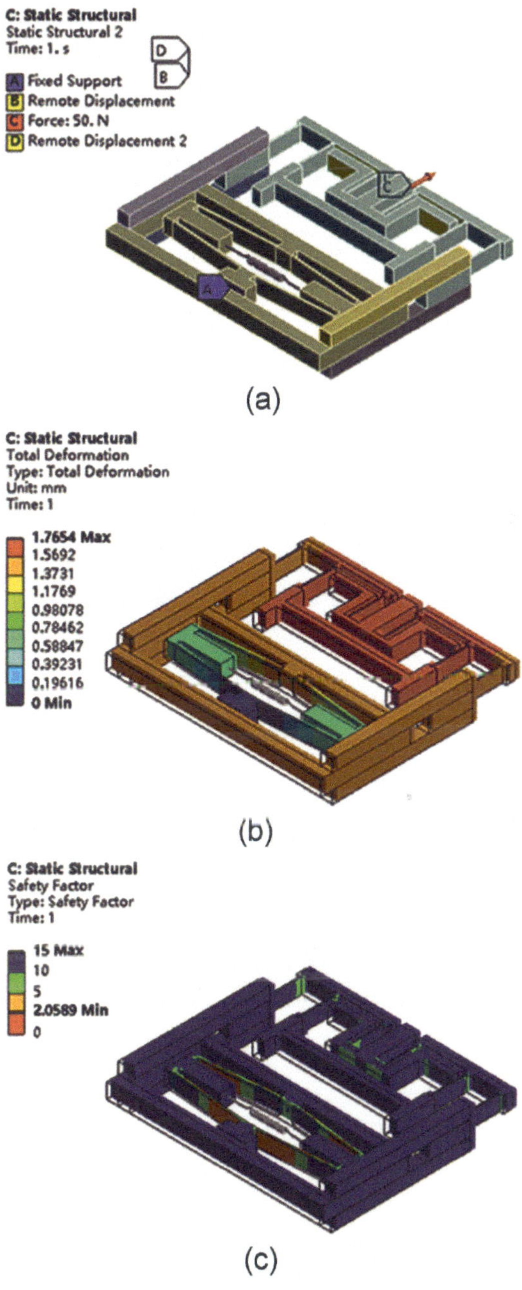

Figure 7.7 Simulation results of mechanism under pull input. (A) Analysis settings; (B) deflection of mechanism; (C) safety factor distribution of mechanism.

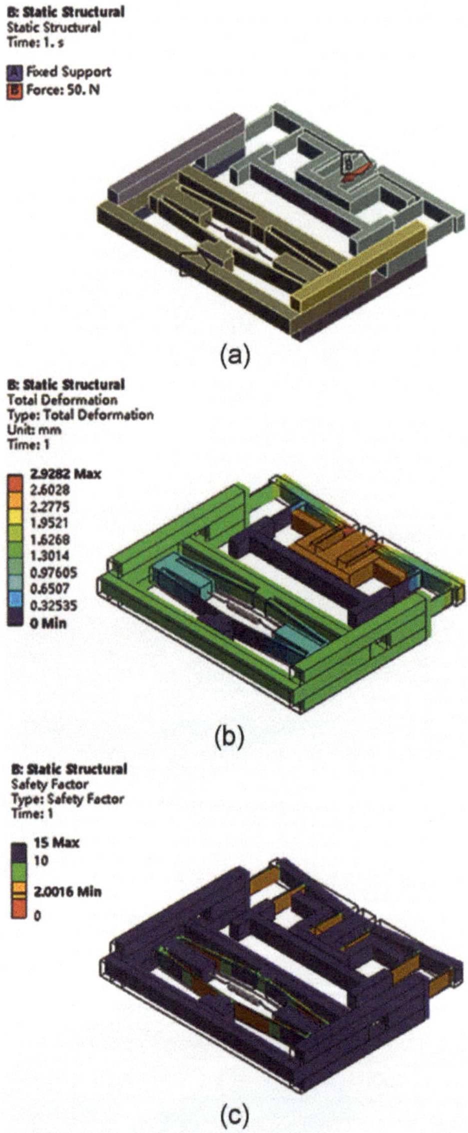

Figure 7.8 Simulation results of mechanism under push input. (A) Analysis settings; (B) deflection of mechanism; (C) safety factor distribution of mechanism.

Fig. 7.9. A voice coil motor (NCC05-18-060-2X, H2W Technologies, Inc.) was used to supply input force for the experiments. Deflection of the flexure mechanism was measured by a laser displacement sensor (LK-H055, Keyence Inc.). The amplitude of the input was measured by a force sensor

Table 7.1 Parameter optimization results of harvester mechanism.

Parameter	Value (mm)	Parameter	Value (mm)
l_1	25.1	t_2	0.5
h_1	1.8	l_{cd}	11.3
t_1	0.6	l_{bd}	16.8
l_2	12.3		

(LSB20, FUTEK Advanced Sensor Technology, Inc.). The measured data were collected by NI PXIe-1082. The load resistance was supplied by a variable resistance box (ZX21A, Shanghai Precision Instruments, Inc.). In addition, a piezoelectric stack (RP150/5x5/30, Harbin Soluble Core Tech Co., Ltd.) was used in the experimental study. The main properties of the piezostack are listed in Table 7.2.

7.4.2 Effect of load resistance

First, the experiment was conducted to verify the effect of load resistance on the average output power of the proposed harvester. The resistance of the load affected the magnitude of output power, and the best resistance of the load depended on the frequency of the input force. Typically, piezoelectric devices are good for high-frequency applications. In low-frequency cases, they do not work optimally. In this work, a piezoelectric energy harvester was designed to harvest energy excited by pull and push inputs, such as humans working on the floor. Thus, low-frequency applications were considered here. For illustration, in the experiment, the proposed harvester was excited by a sinusoidal input force with an amplitude of 10 N

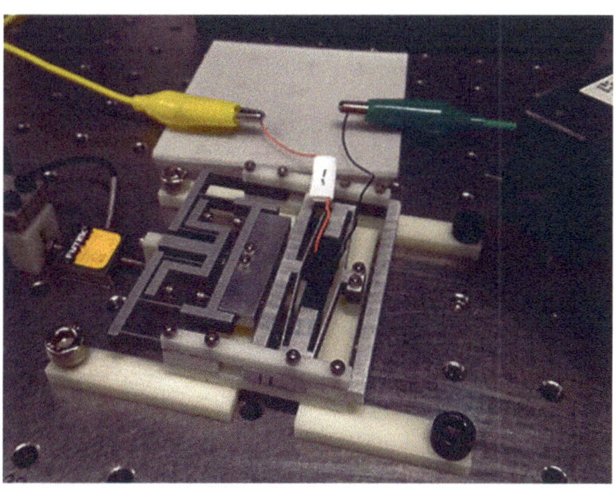

Figure 7.9 Prototype of fabricated energy harvester.

Table 7.2 Main properties of adopted piezoelectric stack.

Property	Value
Dimension	5 mm × 5 mm × 28 mm
Stiffness	38 N/μm
Capacitance	2.00 μF
Coefficient d_{33}	650×10^{-12} C/N

at 5 Hz. The relationship of the output power versus load resistance is shown in Fig. 7.10.

Fig. 7.10 shows that the average output power reached its peak value at 11 kΩ under the pull input, at 12 kΩ under the push input, and at 11 kΩ

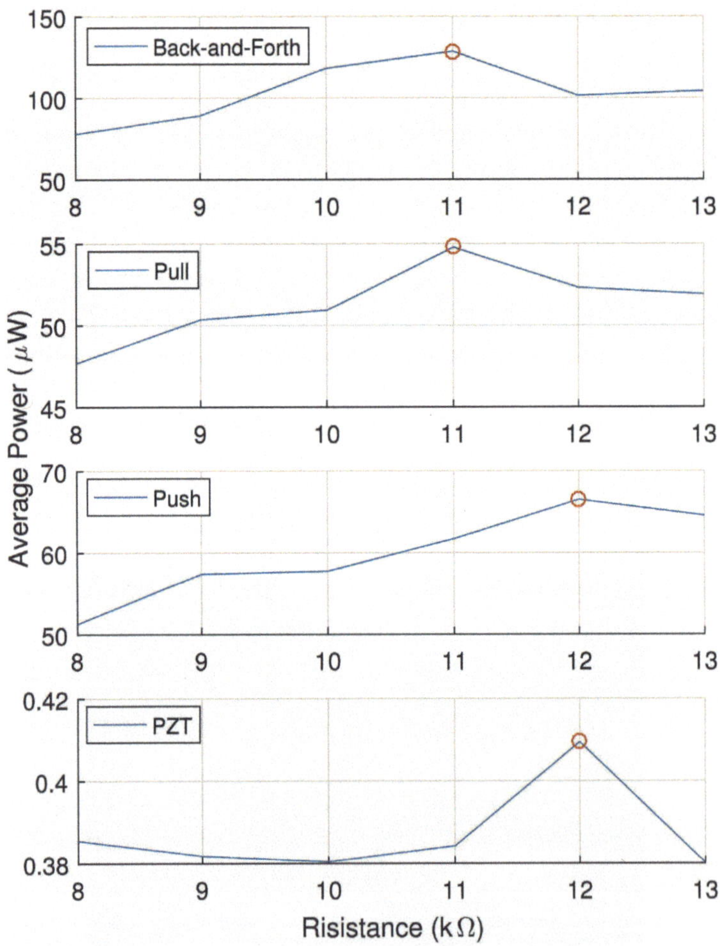

Figure 7.10 Output power with various load resistances under different kinds of sinusoidal input force with an amplitude of 10 N at 5 Hz.

under the back-and-forth input. In addition, when force was directly imposed on the piezoelectric stack, the maximal power was achieved at 12 kΩ. Referring to Eq. (7.20), as the effective piezoelectric constant C_p^{eq} of the piezoelectric stack is 2.4 μF, the optimal resistance at the excitation frequency of 5 Hz can be calculated as 13 kΩ, which is close to the experimental result.

Moreover, the maximum average output power was 0.41 μW when the force acted directly on the piezostack. The maximum average output power was 54.78 μW under pull input, 66.50 μW under push input, and 128.51 μW under back-and-forth input, which was 133.61, 162.20, and 313.44 times the maximum average output power of the piezoelectric stack, respectively. Although the power was relatively low, it was sufficient to supply low-power micro-electromechanical system devices (e.g., wireless sensors), which have ultralow power consumption.

7.4.3 Test with pull input

Next, an experiment was conducted to verify the performance of the proposed harvester under a pull input. The output electric circuit was connected with 11 kΩ resistance, which generated the maximum average output power under a pull input at 5 Hz. The time history of the output voltage under sinusoidal pull input with an amplitude of 10 N at 5 Hz is shown in Fig. 7.11.

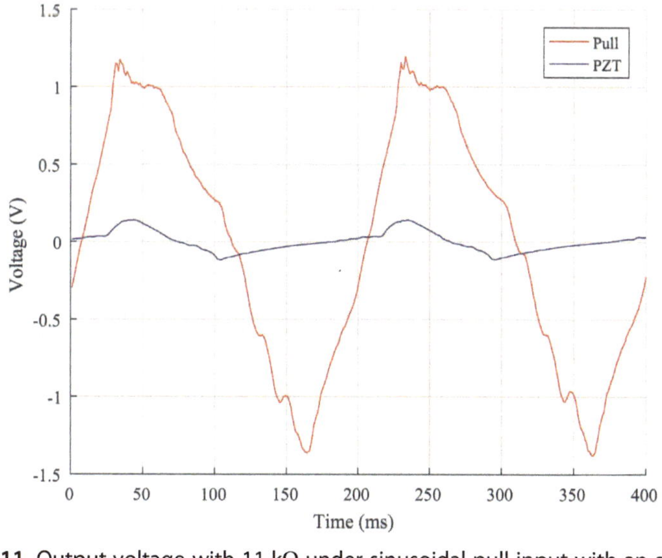

Figure 7.11 Output voltage with 11 kΩ under sinusoidal pull input with an amplitude of 10 N at 5 Hz.

The trend of the curve was similar to the standard unidirectional piezoelectric energy harvester based on a rhombus-type amplifier under a sinusoidal input force (Feenstra et al., 2008; Wang et al., 2016), which verifies that the proposed harvester worked smoothly and the selectivity lever was practicable under the pull input.

Across an optimal resistance of 11 kΩ, the peak—peak output voltage under the pull input reached 2.58 V. In contrast, when the force acted directly on the piezoelectric stack, the peak—peak voltage was only 0.26 V. Therefore, the ratio of the peak—peak voltage of the harvester under the pull input and the peak—peak voltage of the piezostack was calculated as 9.92. Comparing the result with the force amplification ratio under a pull input, it is observed that the error under the pull input is 18.69%.

7.4.4 Test with push input

An experiment is carried out to verify the performance of the proposed harvester under a push input. The output electric circuit was connected with 12 kΩ resistance, which generated the maximum average output power under the push input force at 5 Hz. The time history of the output voltage under sinusoidal push input with an amplitude of 10 N at 5 Hz is shown in Fig. 7.12.

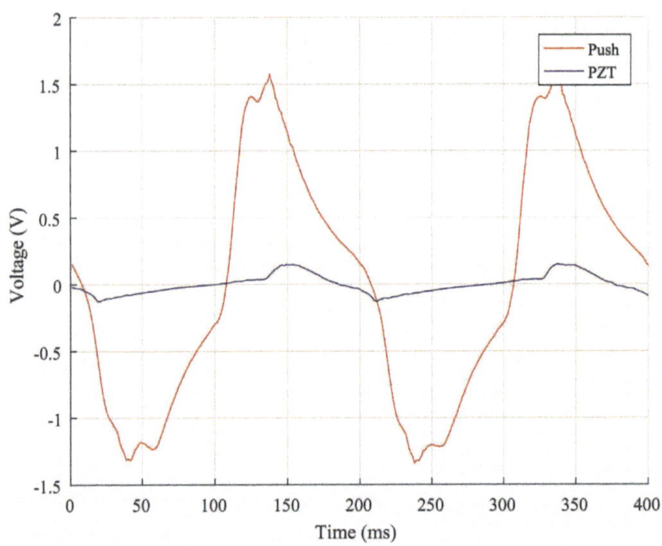

Figure 7.12 Output voltage with 12 kΩ under sinusoidal push input with an amplitude of 10 N at 5 Hz.

The variation trend of the curve was similar to a conventional unidirectional piezoelectric energy harvester based on a rhombus-type amplifier under a sinusoidal input force (Feenstra et al., 2008; Wang et al., 2016), which confirms that the proposed harvester can work smoothly and the selectivity lever is practicable under a push input.

The peak—peak output voltage of the harvester under a push input reached 2.92 V, and the peak-peak output voltage of the stand-alone piezostack was 0.28 V. Therefore, the ratio of the peak—peak voltage of the harvester under push input and the piezostack was calculated as 10.43. Comparing the result with the force amplification ratio under the push input, the error under the push input was around 20.62%.

7.4.5 Test with back-and-forth input

An experiment was conducted to verify the performance of the proposed harvester under a back-and-forth input. The output electric circuit was connected with 11 kΩ resistance, which generated the maximum average output power with an input force at 5 Hz. The time history of the output voltage under sinusoidal back-and-forth input with an amplitude of 10 N at 5 Hz is depicted in Fig. 7.13.

Compared with the curve under only pull or push input, the output of the piezostack under the back-and-forth input doubled, which verifies that

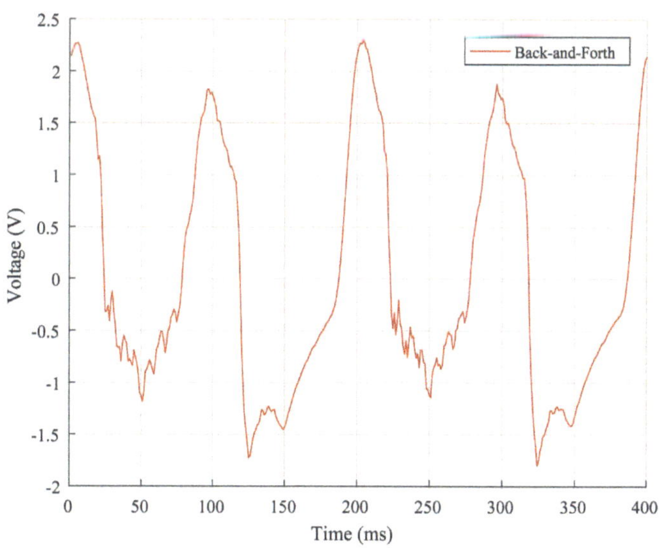

Figure 7.13 Output voltage with 11 kΩ under sinusoidal back-and-forth input with an amplitude of 10 N at 5 Hz.

Table 7.3 Performance comparison with existing design.

Performance	Pull input	Push input	Reference (Feenstra et al., 2008)
Input force	50 N	50 N	60 N
Amplification ratio (simulation)	12.20	13.14	10.9
Amplification ratio (experiment)	9.92	10.43	–

the proposed harvester also worked and the selectivity lever was feasible under continuous back-and-forth input. The curve exhibited slight vibration during the transition from pull to push input, which may have been caused by the collision and separation between the lever mechanism and position limiters.

7.4.6 Comparison with existing work

For a clear comparison, the results are listed in Table 7.3. According to the simulation results, the amplification ratio of the proposed energy harvester with pull and push inputs improved by 11.9% and 20.6%, respectively, with respect to that of the existing work (Feenstra et al., 2008). It is a reasonable result for the situation with a 10 N smaller maximum input force. In addition, there is a discrepancy between the experimental and simulation results. Even so, the proposed energy harvester can obtain much higher energy than that in previous work (Feenstra et al., 2008) in a situation with both pull and push inputs. As a promising application, the energy harvester can be assembled on the floor to scavenge energy excited by human walking, in which bidirectional friction force generated between footsteps and the harvester will be converted into electricity. Moreover, many such devices can be adopted to improve the output power of the energy harvester.

7.5 Conclusion

This chapter presents the design, optimization, and testing of a novel piezoelectric energy harvester with a single piezostack to scavenge energy from excitation with both pull and push inputs. To satisfy actual demands, a flexure mechanism composed of a selectivity lever and rhombus-type amplifier was designed. To obtain the best output performance, the MOGA method based on FEA was adopted to optimize the structural parameters of the flexure mechanism to produce the maximum acting force of a piezoelectric stack under the constraints. A prototype was fabricated and several experiments were conducted to validate the actual performance of the proposed energy harvester. Experimental results verified that the proposed energy harvester was practicable under pull, push, and back-and-forth inputs. Error under pull input and push input is within 18.69% and 20.62%, respectively, which may be caused by fabrication and assembly errors. In future work, the mechanism will be manufactured with higher machining accuracy to improve the performance of the energy harvester. Moreover, a multiaxial energy harvester will be devised dedicated to multiple directional excitation of the input force.

References

Challa, V. R., Prasad, M. G., Shi, Y., & Fisher, F. T. (2008). A vibration energy harvesting device with bidirectional resonance frequency tunability. *Smart Materials and Structures, 17*(1).

Chen, W., Wang, Y., & Deng, W. (2017). Deformable force amplification frame promoting piezoelectric stack energy harvesting: Parametric model, experiments and energy analysis. *Journal of Intelligent Material Systems and Structures, 28*(7), 827–836. https://doi.org/10.1177/1045389X16657426

Fan, K. Q., Chao, F. B., Zhang, J. G., Wang, W. D., & Che, X. H. (2014). Design and experimental verification of a bi-directional nonlinear piezoelectric energy harvester. *Energy Conversion and Management, 86*, 561–567. https://doi.org/10.1016/j.enconman.2014.06.021

Feenstra, J., Granstrom, J., & Sodano, H. (2008). Energy harvesting through a backpack employing a mechanically amplified piezoelectric stack. *Mechanical Systems and Signal Processing, 22*(3), 721–734. https://doi.org/10.1016/j.ymssp.2007.09.015

Lobontiu, N., Paine, J. S. N., Garcia, E., & Goldfarb, M. (2001). Corner-filleted flexure hinges. *Journal of Mechanical Design, Transactions of the ASME, 123*(3), 346–352. https://doi.org/10.1115/1.1372190

Wang, Y., Chen, W., & Guzman, P. (2016). Piezoelectric stack energy harvesting with a force amplification frame: Modeling and experiment. *Journal of Intelligent Material Systems and Structures, 27*(17), 2324–2332. https://doi.org/10.1177/1045389X16629568

Wen, S., Xu, Q., & Zi, B. (2018). Design of a new piezoelectric energy harvester based on compound two-stage force amplification frame. *IEEE Sensors Journal, 18*(10), 3989–4000. https://doi.org/10.1109/JSEN.2018.2820221

Xu, T. B., Siochi, E. J., Kang, J. H., Zuo, L., Zhou, W., Tang, X., & Jiang, X. (2013). Energy harvesting using a PZT ceramic multilayer stack. *Smart Materials and Structures, 22*(6). https://doi.org/10.1088/0964-1726/22/6/065015

Zhao, S., & Erturk, A. (2014). Deterministic and band-limited stochastic energy harvesting from uniaxial excitation of a multilayer piezoelectric stack. *Sensors and Actuators, A: Physical, 214*, 58–65. https://doi.org/10.1016/j.sna.2014.04.019

CHAPTER 8

Design of a two-dimensional energy harvester with a single piezoelectric stack

This chapter presents the design, fabrication, and testing of a piezoelectric energy harvester. Its unique feature is that a single piezostack is used to harvest energy from two-degree-of-freedom (2-DOF) excitation in the vertical direction. To accept input forces in two directions, a flexure-based two-stage force amplifier (TSFA) is integrated with a conical transmission mechanism (CTM) and a sliding guiding mechanism (SGM). An analytical model is established to predict the force amplification ratio of the proposed energy harvester. To maximize the output performance of the energy harvester, the dominant parameters of the mechanism are determined, which are then optimized by resorting to the multiobjective genetic algorithm (MOGA). A prototype energy harvester is fabricated for experimental testing. Results demonstrate the satisfactory performance of the developed energy harvester.

8.1 Introduction

A lot of piezoelectric stack-based energy harvesters have been reported (Chen et al., 2017; Wang et al., 2017; Wang et al., 2016; Wen et al., 2018). However, most available energy harvesters can scavenge the energy only from a unidirectional input force. Such devices cannot scavenge energy produced in other excitation directions, and that energy is wasted. An energy harvester is more realistic if it can accept input excitation in more directions. To scavenge energy from multiple directions, although multiple 1-DOF piezoelectric harvesters can simply be combined, the piezoelectric materials do not work at the same time for each directional excitation. This results in low power density. In practice, it is more economical to employ an energy harvester with fewer piezoelectric stacks. Therefore, it is desirable to develop a piezoelectric energy harvester with a single piezoelectric stack to scavenge energy from multiple-DOF excitation.

Mechanical Design of Piezoelectric Energy Harvesters
ISBN 978-0-12-823364-1
https://doi.org/10.1016/B978-0-12-823364-1.00011-0
© 2022 Elsevier Inc.
All rights reserved.

Several 2-DOF piezoelectric energy harvesters have been reported (Hu et al., 2018; Wang and Tang, 2017; Xiao et al., 2015). However, these 2-DOF harvesters can accept excitation in only a single direction. Few multiple-DOF energy harvesters have been reported to receive multidirectional excitation by a single piezoelectric stack. The major challenge is how to transmit multidirectional force into a unidirectional force for exertion on the input ends of the single piezoelectric stack material.

This chapter reports the design and development of a new 2-DOF piezoelectric energy harvester with a single piezoelectric stack to scavenge energy from excitation in two directions. In addition to analytical modeling and design optimization, a prototype energy harvester was fabricated for experimental verification of the proposed conceptual design.

8.2 Mechanism design

In this section, the mechanism design of the proposed energy harvester is introduced in detail. The computer-aided design (CAD) model of the energy harvester is shown in Fig. 8.1. To facilitate analysis, assign F_{input} as the input force and F_{PZT} (denoted by F_{30x} and F_{40x}) as the output force.

8.2.1 Working principle

The working principle of this design is explained in the following. As shown in Fig. 8.1B, a top input plate is connected to two CTMs with conical grooves for coupling. The two lug couplings of the CTM fixed onto the two input ends of the flexure-based TSFA. A piezoelectric stack is mounted in the center of the force amplifier. Moreover, each CTM with a conical groove is coupled to the base through an SGM.

When an input force (with y-axis and z-axis components) is applied to the energy harvester, a reaction force perpendicular to the sliding surface of the CTM is generated. Then, the SGM eliminates the y-axis and z-axis components of the reaction force. Meanwhile, the x-axis component of the reaction force is transmitted to the TSFA. In this way, the input force with y-axis and z-axis components is transmitted to the piezoelectric stack that is mounted inside the TSFA. The force amplifier contributes to improving output power to the piezoelectric energy harvester (Xu et al., 2013).

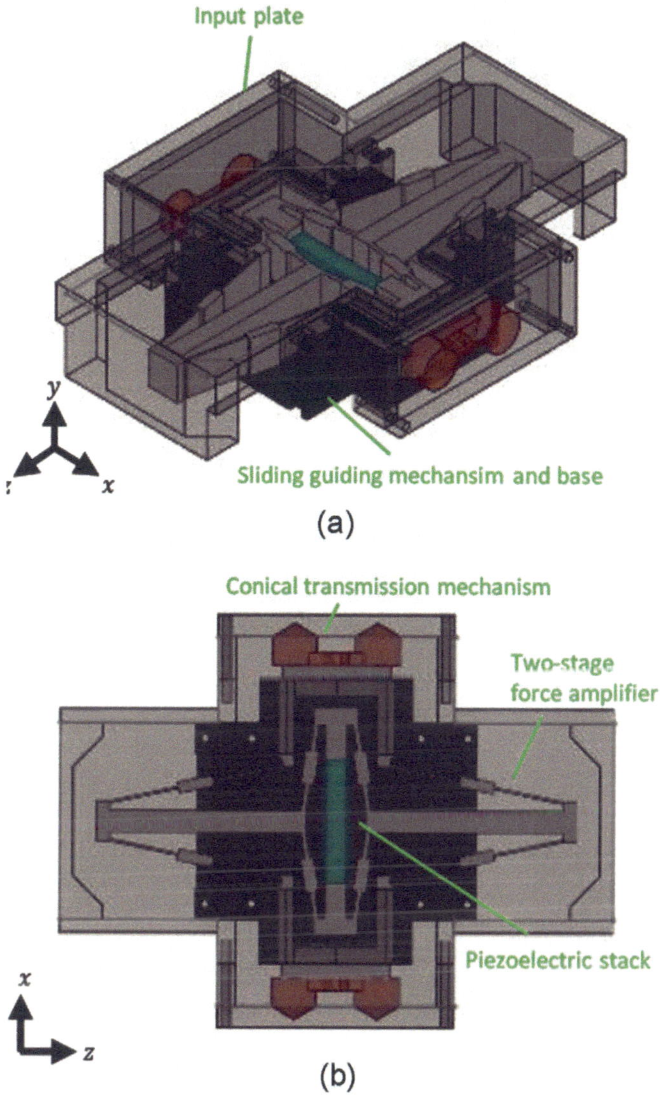

Figure 8.1 Computer-aided design model of the designed energy harvester. (A) Three-dimensional view; (B) Top view.

8.2.2 Model of conical transmission mechanism

In this section, an analytical model is derived to describe the force relationship of the CTM. The static model of the CTM is shown in Fig. 8.2.

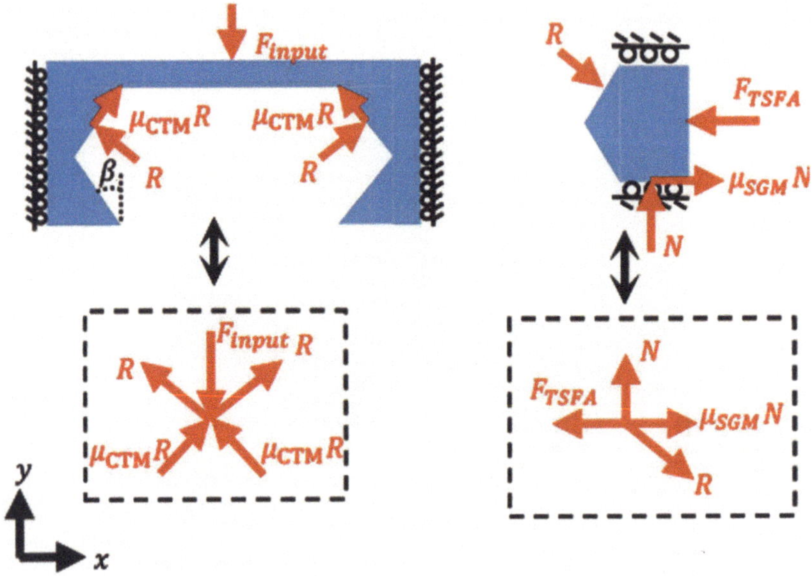

Figure 8.2 Schematic diagram of conical transmission mechanism.

For the static equilibrium, the governing equations of the CTM are determined as:

$$2R\sin\beta - 2\mu_{\mathrm{CTM}}R\cos\beta - F_{\mathrm{input}} = 0 \tag{8.1}$$

$$\mu_{\mathrm{SGM}}N + R\cos\beta - F_{\mathrm{TSFA}} = 0 \tag{8.2}$$

$$N - R\sin\beta = 0 \tag{8.3}$$

where R is the reaction force generated by the inclined plane, β is the angle between the inclined plane and z-axis, F_{TSFA} is the force produced by the TSFA, N is the reaction force generated by the SGM, and μ_{CTM} and μ_{SGM} are the friction coefficients of the sliding surfaces of the CTM and SGM, respectively.

In Eqs. (8.1)–(8.3), the force amplification ratio of the CTM (η_{CTM}) can be derived as:

$$\eta_{\mathrm{CTM}} = \frac{F_{\mathrm{TSFA}}}{F_{\mathrm{input}}} = \frac{\cos\beta + \mu_{\mathrm{SGM}}\sin\beta}{2\sin\beta - 2\mu_{\mathrm{CTM}}\cos\beta} \tag{8.4}$$

Because it is difficult to predict the friction coefficients, all sliding surfaces are assumed to be smooth, so that the friction coefficients are considered to be zero. Therefore, the force amplification ratio (η_{CTM},0) of the CTM without friction coefficient can be derived as:

$$\eta_{\mathrm{CTM},0} = \frac{\cos\beta}{2\sin\beta} = \frac{1}{2}\cot\beta \tag{8.5}$$

8.2.3 Model of two-stage force amplifier

In addition, an analytical model is established to describe the force–deformation relationship of the TSFA. All connecting linkages are regarded as flexure hinges with a small deformation. For each flexure hinge, there are two associated nodes. In this way, a total of 44 nodes are assigned for the TSFA, as shown in Fig. 8.3. The input nodes are represented by numbers 1 and 13, and the output ones are denoted as 30 and 40.

According to Hooke's law, the governing equation of the stage can be expressed as:

$$F = KU \tag{8.6}$$

where K is the global stiffness matrix, and F and U are the external force and displacement for all nodes, which can be expressed as:

$$F = \left[(F_1)^T \quad \cdots \quad (F_i)^T \quad \cdots \quad (F_n)^T \right] \tag{8.7}$$

$$U = \left[(U_1)^T \quad \cdots \quad (U_i)^T \quad \cdots \quad (U_n)^T \right] \tag{8.8}$$

where n is the total number of nodes, and F_i and U_i are the force and displacement vectors of node i, respectively.

In addition, F_i and U_i can be written in the following forms:

$$F_i = \left[F_{ix} \quad F_{iz} \quad M_{iy} \right], \ i = 1, 2, \cdots, n \tag{8.9}$$

$$U_i = \left[x_i \quad z_i \quad \theta_{iy} \right], \ i = 1, 2, \cdots, n \tag{8.10}$$

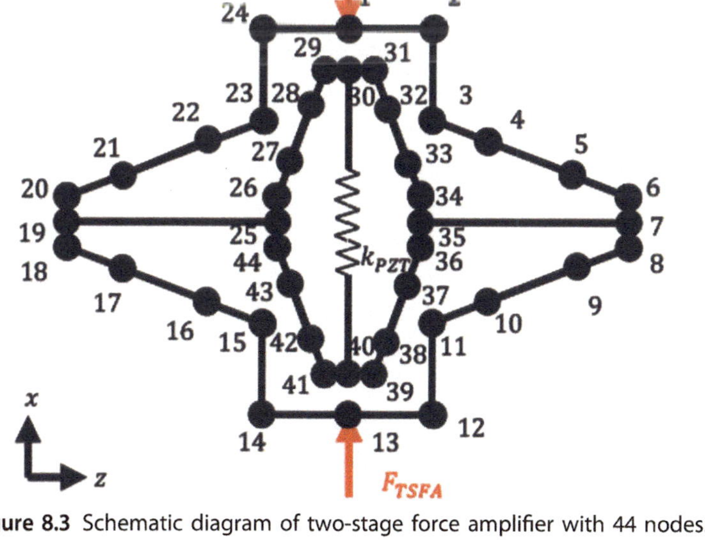

Figure 8.3 Schematic diagram of two-stage force amplifier with 44 nodes.

where F_{ix}, F_{iz}, and M_{iy} are the force components of node i; and x_i, z_i, and θ_{iy} are the displacement components of node i.

In Fig. 8.4, the governing equations of flexure hinge ij in local coordinates can be derived as:

$$F_i = K_{ij}^{11} U_a + K_{ij}^{12} U_b \tag{8.11}$$

$$F_j = K_{ij}^{21} U_a + K_{ij}^{22} U_b \tag{8.12}$$

with $K_{ij}^{11} = K_{ij}^{22}$ and $K_{ij}^{12} = \left(K_{ij}^{21}\right)^T$, in which K_{ij}^{11} and K_{ij}^{12} are expressed by:

$$K_{ij}^{11} = \begin{bmatrix} \frac{EA}{l} & 0 & 0 \\ 0 & \frac{12EI}{l^3} & \frac{6EI}{l^2} \\ 0 & \frac{6EI}{l^2} & \frac{4EI}{l} \end{bmatrix} \tag{8.13}$$

$$K_{ij}^{12} = \begin{bmatrix} \frac{EA}{l} & 0 & 0 \\ 0 & -\frac{12EI}{l^3} & \frac{6EI}{l^2} \\ 0 & -\frac{6EI}{l^2} & \frac{2EI}{l} \end{bmatrix} \tag{8.14}$$

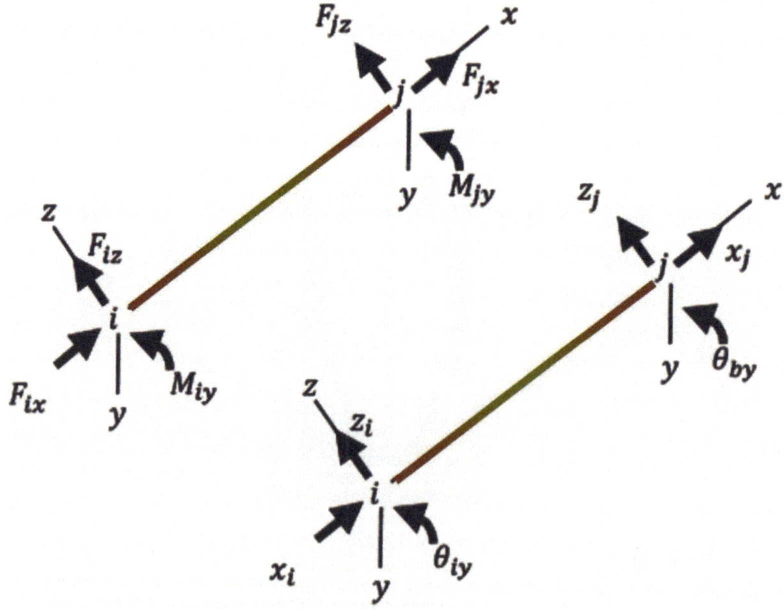

Figure 8.4 Schematic diagram of flexure hinge ij in local coordinates.

where E is the Young's modulus; and l, A, and I are the length, cross-section area, and moment of inertia of flexure hinge ij, respectively.

Then, local element stiffness matrix (K_{ij}^L) of flexure hinge ij can be derived as:

$$K_{ij}^L = \begin{bmatrix} K_{ij}^{11} & K_{ij}^{12} \\ K_{ij}^{21} & K_{ij}^{22} \end{bmatrix} \tag{8.15}$$

To transform it into a global element stiffness matrix, the rotational transformation matrix is given as:

$$T_{ab}(\alpha) = \begin{bmatrix} \cos\alpha & \sin\alpha & 0 & 0 & 0 & 0 \\ -\sin\alpha & \cos\alpha & 0 & 0 & 0 & 0 \\ 0 & 0 & 1 & 0 & 0 & 0 \\ 0 & 0 & 0 & \cos\alpha & \sin\alpha & 0 \\ 0 & 0 & 0 & -\sin\alpha & \cos\alpha & 0 \\ 0 & 0 & 0 & 0 & 0 & 1 \end{bmatrix} \tag{8.16}$$

where α is the rotation angle (in a counterclockwise direction) from the local coordinate to the global coordinate.

Hence, global element stiffness matrix K_{ab} can be obtained as:

$$K_{ab} = T_{ab}(\alpha) K_{ab}^L [T_{ab}(\alpha)]^T \tag{8.17}$$

The boundary conditions are assigned according to the essential and natural boundary limits. The input forces are applied at nodes 1 and 13, with $F_1 = [-|F_{\text{TSFA}}| \ \ 0 \ \ 0]^T$ and $F_{13} = [|F_{\text{TSFA}}| \ \ 0 \ \ 0]^T$. Moreover, the output forces are applied at nodes 30 and 40, with $F_{30} = [|K_{\text{PZT}}(x_{40} - x_{30})| \ \ 0 \ \ 0]^T$ and $F_{40} = [-|K_{\text{PZT}}(x_{40} - x_{30})| \ \ 0 \ \ 0]^T$, respectively.

Therefore, the force amplification ratio (η_{TSFA}) of the TSFA is given by:

$$\eta_{\text{TSFA}} = \frac{|F_{30x}|}{|F_{1x}|} \tag{8.18}$$

According to Eqs. (8.5) and (8.18), the total force amplification ratio (η_{total}) of the proposed harvester is calculated as:

$$\eta_{total} = \eta_{CTM,0} \cdot \eta_{TSFA} \tag{8.19}$$

8.2.4 Model of piezoelectric stack

Assume that harmonic force $F(t) = F_0 e^{-j\omega t}$ is exerted on the piezoelectric stack, where F_0 is the amplitude of the force. Induced complex voltage (V) can be calculated as (Zhao and Erturk, 2014):

$$V = \frac{j\omega d_{33}^{eff} F_0}{j\omega C_p^{eq} + (1/R_l)} \tag{8.20}$$

where d_{33}^{eff} is the effective piezoelectric constant, C_p^{eq} is the equivalent capacitance, and R_l is the resistance of load, respectively.

In Eq. (8.20), the open-circuit voltage (V_{open}) and output power (P) can be derived as:

$$V_{open} = \frac{d_{33}^{eff} F_0}{C_p^{eq}} \tag{8.21}$$

$$P = \frac{|V|^2}{R_l} = \frac{\left(\omega d_{33}^{eff} F_0\right)^2}{\left(\omega C_p^{eq}\right)^2 R_l + 1/R_l} \tag{8.22}$$

According to Eq. (8.22), the output power can be maximized when $R_l = 1 \big/ \left(\omega C_p^{eq}\right)$. Thus, the maximum output power (P_{max}) can be calculated as:

$$P_{max} = \frac{\omega \left(d_{33}^{eff} F_0\right)^2}{2 C_p^{eq}} \tag{8.23}$$

8.3 Parameter optimization

In this section, the parameters of the proposed energy harvester mechanism are optimized by using the MOGA method based on finite element analysis (FEA) simulation. The selection of variable parameters, constraints of the FEA, and results of optimization are presented next.

8.3.1 Variable parameter selection

Before optimization, the dominant parameters that mostly influence the performance of the proposed energy harvester should be determined. In Eq. (8.5), angle β has a significant impact on the performance of the CTM. It indicates that the force amplification ratio (η_{CTM}) of the CTM is increased when angle β becomes smaller. However, considering the reliability of operation and manufacturing, $\beta = 30$ degrees is selected. It provides a force amplification ratio of $\eta_{CTM} = 0.866$ for the CTM.

In addition, length (L), cross-section area (A), and rotation angle (α) (i.e., counterclockwise angle from the local to global coordinate) of the flexure hinges are key parameters that most influence the performance of the T_{SFA}. To facilitate optimization, some parameters of the thick beam are fixed. Thus, the variable parameters are determined as:

$$[x_{3\cdot6}, z_{3\cdot6}, t_{3\cdot4}, z_{3\cdot4}, x_{26\cdot29}, z_{26\cdot29}, t_{26\cdot27}, x_{26\cdot27}, d] \quad (8.24)$$

where parameters $x_{i\cdot j}$, $z_{i\cdot j}$, $t_{i\cdot j}$, and d are denoted in Fig. 8.5.

8.3.2 Optimization setup and results

In this work, the MOGA method based on FEA simulation (with ANSYS Workbench software) is adopted for optimization. Before optimization is implemented, the model of the mechanism is parameterized to make the structure change based on the defined parameter relationships. In addition, the material of the mechanism is selected as Al-7075 owing to its high elasticity.

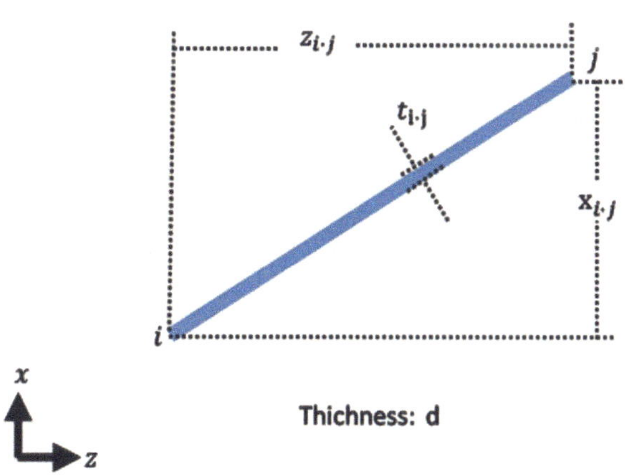

Figure 8.5 Illustration of main parameters for flexure hinge *ij*.

First, the static structural analysis of the TSFA is conducted. To facilitate a comparison of the performance of the designed harvester with those in the literature, the magnitude of the input force (F_{TSFA}) is selected as 100 N for the TSFA. Then, the input force (F_{input}) of the proposed harvester can be derived as 115.47 N. Because the stiffness of the piezostack is much larger, both ends of the piezostack are treated as fixed supports. Thus, the output force is obtained by detecting the value of the reaction force for the fixed support. The analysis settings for the simulation study are shown in Fig. 8.6(A).

To generate higher output energy, the key objective of optimization is to maximize the magnitude of the output force in the static structural analysis. In addition, to avoid material failure, it is necessary to set a constraint on the safety factor (greater than 1.0) for the adopted material. In the case study, the safety factor is assigned to be greater than 2.0. To facilitate manufacturing and optimization, the grid sizes of $x_{i\cdot j}$, $z_{i\cdot j}$, and $t_{i\cdot j}$ are assigned as 0.1 mm, and the grid of d is chosen to be 1 mm.

Based on these constraints, the optimization results of the variable parameters are generated as shown in Table 8.1. Under an input force of 100 N, the output force is obtained as 2398 N for the TSFA. The force amplification ratios of the TSFA and the proposed harvester are derived as $\eta_{TSFA} = 23.98$ and $\eta_{total} = 20.77$, respectively. The simulation result of deformation for the TSFA is depicted in Fig. 8.6(B). It indicates that the maximum deflection of the input nodes for the TSFA is 2.02 mm. Therefore, the height of the cone for the CTM is assigned as 3 mm, and the maximum displacement of the input plate is estimated to be 3.50 mm. The distribution of safety factor values for the TSFA is shown in Fig. 8.6(C), which reveals that the minimum safety factor of the TSFA is 2.002, which satisfies the imposed constraints for optimization.

8.4 Prototype fabrication and experimental results

In this section, a prototype energy harvester with optimized structural parameters is fabricated to verify the performance of the designed harvester.

8.4.1 Prototype fabrication and experimental setup

Based on the CAD model, the TSFA is manufactured using Al-7075 material with a tolerance of ± 0.05 mm. Other components are designed to satisfy the functional demands of the proposed design, especially to reduce the friction coefficients of the sliding surfaces. These components are manufactured by three-dimensional printing.

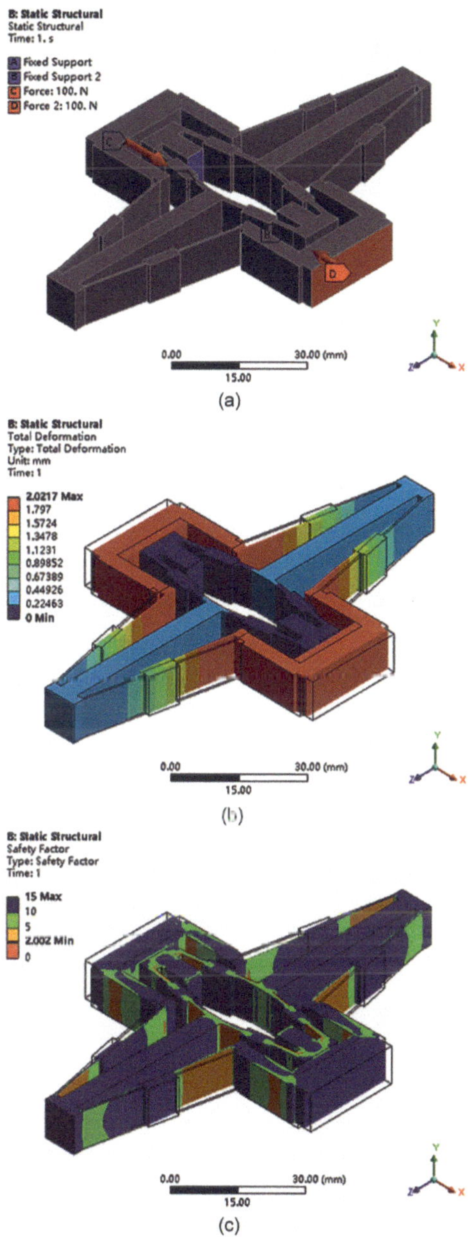

Figure 8.6 Simulation results of the two-stage force amplifier. (A) Analysis setting; (B) Deformation result; (C) Safety factor distribution.

Table 8.1 Optimization results of harvester parameters.

Parameter	Value	Parameter	Value
$x_{3\cdot 6}$	8.2 mm	$z_{26\cdot 29}$	3 mm
$z_{3\cdot 6}$	40 mm	$t_{26\cdot 27}$	1.2 mm
$t_{3\cdot 4}$	0.5 mm	$x_{26\cdot 27}$	5.9 mm
$z_{3\cdot 4}$	16.1 mm	d	10 mm
$x_{26\cdot 29}$	19.3 mm		

The fabricated prototype and experimental setup are shown in Fig. 8.7. A voice coil motor (NCC05-18-060-2X, H2W Technologies, Inc.) is adopted to provide the input force for the experimental setup. A laser displacement sensor (LK-H055, Keyence Inc.) is used to measure the deflection displacement of the flexure mechanism. In addition, a force sensor (LSB20, FUTEK Advanced Sensor Technology, Inc.) is used to measure the input force. A variable resistance box (ZX21A, Shanghai Precision Instruments Inc.) is employed to supply the load resistance. The measured data are collected by an NI PXIe-1082 real-time controller. Also, a piezoelectric stack (RP150/5*5/30, Harbin Soluble Core Tech Co., Ltd.) is chosen for the experimental study. The main properties of the adopted piezoelectric stack are listed in Table 8.2.

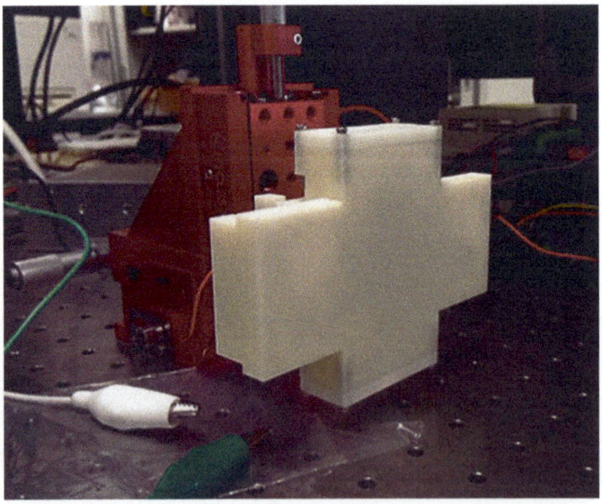

Figure 8.7 Fabricated prototype harvester and experimental setup.

Table 8.2 Main parameters of adopted piezoelectric stack.

Property	Symbol	Value
Dimension	$a \times a \times l$	$5 \times 5 \times 28$ mm
Stiffness	k_{PZT}	48.26 N/μm
Capacitance	C_p^{eq}	2.40 μF
Coefficient	d_{33}	650×10^{-12} C/N

8.4.2 Influence of load resistance

First, an experimental study was conducted to verify the influence of load resistance on the average output power of the developed harvester. For illustration, the proposed harvester was excited by a sinusoidal input force with an amplitude of 10 N at 5 Hz. The relationship of the output power versus load resistance is shown in Fig. 8.8.

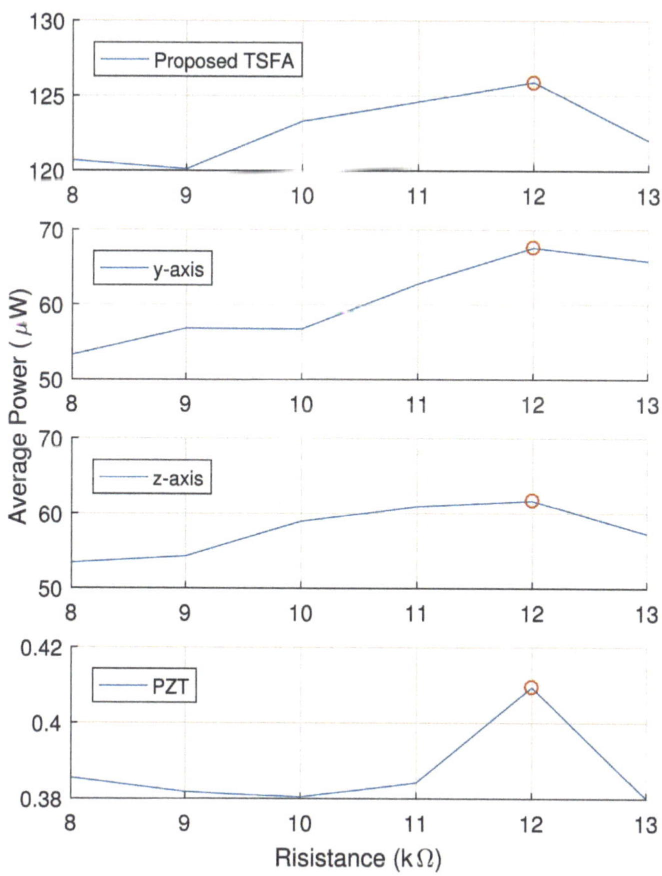

Figure 8.8 Experimental results of output power with various load resistances under sinusoidal input force with an amplitude of 10 N at 5 Hz.

The average output power reached its peak value at 12 kΩ with the input force exerted on the proposed TSFA. That is, the input force was applied on the y-axis or z-axis. In addition, when the force was directly applied on the piezoelectric stack, the maximal power was also achieved at 12 kΩ. When the effective piezoelectric constant Cpeq of the piezoelectric stack was 2.4 μF, optimal resistance at an excitation frequency of 5 Hz was calculated to be 13 kΩ, which is close to the experimental result.

Moreover, when force was directly exerted on the piezostack, the maximum average output power was 0.41 μW. By contrast, with the same force acting on the proposed TSFA, the maximum average output power was 125.9 μW, which is 307.07 times higher than that of the stand-alone piezostack. In addition, the maximum average output power was 67.58 and 61.69 μW when the input force was applied on the y-axis and z-axis, which was 164.83 and 150.46 times the maximum average output power of the piezostack, respectively. The obtained electric power was sufficient to supply low-power micro-electromechanical system devices (e.g., wireless sensors), which have ultralow power consumption.

8.4.3 Performance test of two-stage force amplifier

Next, an experimental study was carried out to test the performance of the proposed TSFA. The output electric circuit was connected with 12-kΩ resistance, which generated the maximum average output power under the input force directly acting on the proposed TSFA at 5 Hz. Under a sinusoidal input force with an amplitude of 10 N at 5 Hz, the time history of the output voltage was obtained, as shown in Fig. 8.9.

The peak—peak output voltage under the input force reached 6.272 V. In contrast, when the force was directly applied to the piezoelectric stack, the peak—peak voltage was only 0.28 V. Therefore, the ratio of the peak—peak voltages of the TSFA and the piezostack was calculated to be 22.4. In a comparison of the result with the calculated force amplification ratio, the experimental result was 6.59% lower than the model result. The discrepancy mainly arose from fabrication error of the mechanism parameters and simplification adopted in modeling.

8.4.4 Performance test of proposed energy harvester

Next, an experimental study was conducted to verify the performance of the proposed harvester. The output electric circuit was connected with a 12-kΩ resistance, which generated the maximum average output power

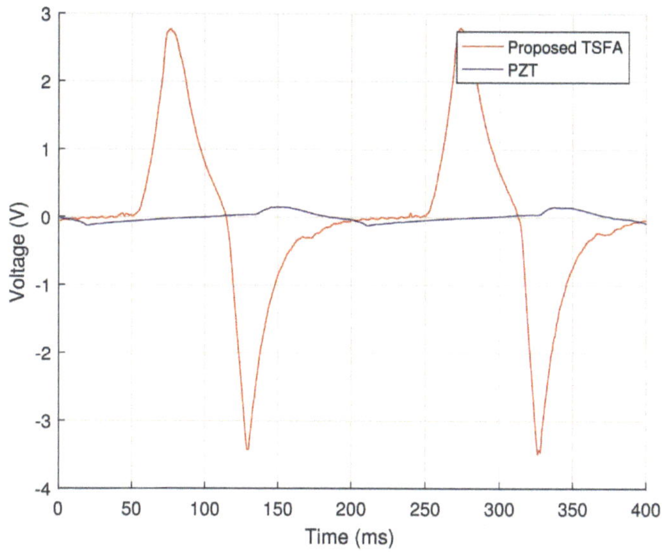

Figure 8.9 Output voltages of two-stage force amplifier and piezostack at 12 kΩ under a sinusoidal input force with an amplitude of 10 N at 5 Hz.

under input forces in the y-axis and z-axis at 5 Hz. Under both sinusoidal input forces in the y-axis and z-axis with an amplitude of 10 N at 5 Hz, the time histories of the output voltage were generated, as shown in Fig. 8.10.

The variation tendency of the curve was similar to that of the standard piezoelectric energy harvester under a sinusoidal input force (Wang et al., 2016). This confirms that the proposed harvester worked smoothly and the CTM was feasible under 2-DOF input force. The curve exhibited slight vibration, which may have been caused by friction generated from the CTM and SGM.

The peak−peak output voltages under the input force in the y-axis and z-axis reached 5.34 and 4.66 V, respectively. In contrast, when the force was directly applied to the piezoelectric stack, the peak−peak voltage was only 0.28 V. Therefore, the ratios of the peak−peak voltages of the harvester and the piezostack were calculated to be 19.07 and 16.64 under the input force in the y-axis and z-axis, respectively. In a comparison of results with the ideal total force amplification ratio, the deviations were 8.18% and 16.64% under the input force in the y-axis and z-axis, respectively.

Afterward, the force amplification ratios of the CTM in the y-axis and z-axis were derived as 0.852 and 0.743, respectively, which were 1.6% and

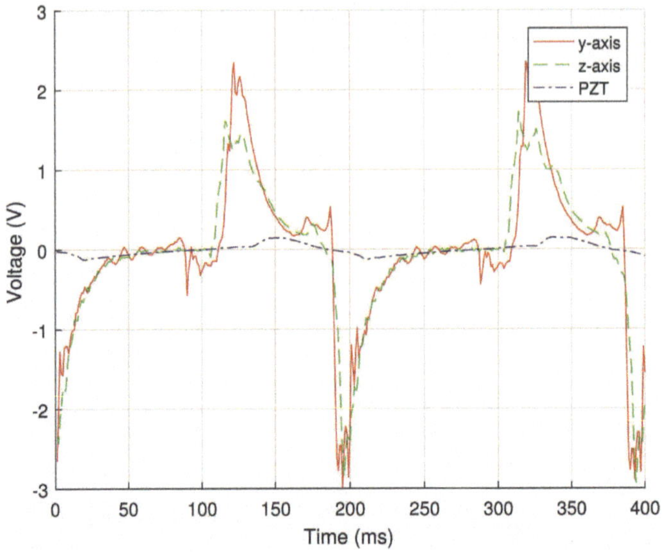

Figure 8.10 Experimental results of output voltage of energy harvester with load resistance of 12 kΩ under a sinusoidal input force in y-axis and z-axis with an amplitude of 10 N at 5 Hz.

14.2% smaller than the calculated value of 0.866. The discrepancy of the force amplification ratio for the CTM may have been caused by the friction coefficient of the mechanism.

8.4.5 Comparison study versus existing work

For a clear comparison study, the performances of typical force amplifiers are listed in Table 8.3. According to the simulation results, both the force amplification ratio and safety factor of the proposed TSFA were improved compared with those of the existing work (Wang et al., 2017). In addition, there was a discrepancy between the experimental and simulation results. Even so, the force amplification ratio of the proposed TSFA was larger than that of the previous work (Wang et al., 2017). Moreover, for the simulation result of the proposed energy harvester, the total force amplification ratio was 20.77, which is close to that of the previous work (Wang et al., 2017). However, for the experimental result, the total force amplification ratios under the input forces in the y-axis and z-axis were 9.2% and 20.8% smaller than those of the previous work (Wang et al., 2017). However, the proposed energy harvester can obtain much higher energy than that in the previous work (Wang et al., 2017) under 2-DOF excitation.

Table 8.3 Performance comparison of designed harvester with existing work.

Performance	Proposed harvester (y-axis)	Proposed harvester (z-axis)	Proposed two-stage force amplifier	Previous two-stage force amplifier in (Wang et al., 2017)
Safety factor	2.002	2.002	2.002	1.6
Input force	115.47 N	115.47 N	100 N	100 N
Amplification ratio (simulation)	20.77	20.77	23.98	21
Amplification ratio (experiment)	19.07	16.64	22.4	21

8.5 Conclusion

This chapter presents the design, optimization, fabrication, and testing of a new piezoelectric energy harvester with a single piezostack to scavenge energy from excitation with a 2-DOF direction. To satisfy actual demand, a flexure-based mechanism composed of TSFA, CTM, and SGM was designed. To achieve the best output performance, the MOGA method based on FEA was implemented to optimize the structural parameters of the flexure mechanism, which produced the maximum acting force on the piezostack under the constraints. A prototype energy harvester was fabricated and several experiments were conducted to demonstrate the performance of the developed harvester. Experimental results verified that the proposed energy harvester was feasible under 2-DOF input force. It provided the maximum average output power of 67.58 and 61.69 μW for 10-N input in the y-axis and z-axis, respectively, which is sufficient for low-power electronic devices. In the future, the mechanism will be manufactured with higher machining accuracy to improve the performance of the energy harvester.

References

Chen, W., Wang, Y., & Deng, W. (2017). Deformable force amplification frame promoting piezoelectric stack energy harvesting: Parametric model, experiments and energy analysis. *Journal of Intelligent Material Systems and Structures, 28*(7), 827–836. https://doi.org/10.1177/1045389X16657426

Hu, G., Tang, L., Das, R., & Marzocca, P. (2018). A two-degree-of-freedom piezoelectric energy harvester with stoppers for achieving enhanced performance. *International Journal of Mechanical Sciences, 149*, 500–507. https://doi.org/10.1016/j.ijmecsci.2017.07.051

Wang, H., & Tang, L. (2017). Modeling and experiment of bistable two-degree-of-freedom energy harvester with magnetic coupling. *Mechanical Systems and Signal Processing, 86*, 29–39. https://doi.org/10.1016/j.ymssp.2016.10.001

Wang, Y., Chen, W., & Guzman, P. (2016). Piezoelectric stack energy harvesting with a force amplification frame: Modeling and experiment. *Journal of Intelligent Material Systems and Structures, 27*(17), 2324–2332. https://doi.org/10.1177/1045389X16629568

Wang, L., Chen, S., Zhou, W., Xu, T. B., & Zuo, L. (2017). Piezoelectric vibration energy harvester with two-stage force amplification. *Journal of Intelligent Material Systems and Structures, 28*(9), 1175–1187. https://doi.org/10.1177/1045389X16667551

Wen, S., Xu, Q., & Zi, B. (2018). Design of a new piezoelectric energy harvester based on compound two-stage force amplification frame. *IEEE Sensors Journal, 18*(10), 3989–4000. https://doi.org/10.1109/JSEN.2018.2820221

Xiao, H., Wang, X., & John, S. (2015). A dimensionless analysis of a 2DOF piezoelectric vibration energy harvester. *Mechanical Systems and Signal Processing, 58*, 355–375. https://doi.org/10.1016/j.ymssp.2014.12.008

Xu, T. B., Siochi, E. J., Kang, J. H., Zuo, L., Zhou, W., Tang, X., & Jiang, X. (2013). Energy harvesting using a PZT ceramic multilayer stack. *Smart Materials and Structures, 22*(6). https://doi.org/10.1088/0964-1726/22/6/065015

Zhao, S., & Erturk, A. (2014). Deterministic and band-limited stochastic energy harvesting from uniaxial excitation of a multilayer piezoelectric stack. *Sensors and Actuators A: Physical, 214*, 58–65. https://doi.org/10.1016/j.sna.2014.04.019

CHAPTER 9

Design of a two-dimensional piezoelectric energy harvester with magnets and multistage force amplifier

This chapter presents the design and testing of a two-directional (2-D) piezoelectric stack-based energy harvester to capture mechanical energy from human footsteps. The energy of 2-D excitation is scavenged by a single piezoelectric stack, which improves power density. The harvester is constructed by integrating permanent magnets, position protection plates, and a multistage force amplification frame. Through analytical modeling, the main architectural parameters of the device are determined. A finite element analysis (FEA) simulation study is conducted to optimize device parameters to achieve the largest force amplification ratio with a high safety factor. A prototype harvester is fabricated for experimental study. Results demonstrate the satisfactory performance of the developed 2-D piezo energy harvester.

9.1 Introduction

To generate a larger output power for a piezoelectric stack-based energy harvester, a force amplifier mechanism is beneficial to magnify force exerted on the piezoelectric stack. Investigations on enhancing the output performance of the force amplification frame have been carried out extensively (Chen et al., 2017; Feenstra et al., 2008; Wang et al., 2017; Wang et al., 2016). Nevertheless, most previous work focused only on improving the force amplification ratio and energy conversion efficiency. Researchers were less concerned with the practical application issue of devices. For instance, in previous work (Wen et al., 2018), a two-stage force amplification frame with compound beams was proposed to achieve a large safety factor and force amplification ratio. However, it is lack of mechanism

Mechanical Design of Piezoelectric Energy Harvesters
ISBN 978-0-12-823364-1
https://doi.org/10.1016/B978-0-12-823364-1.00008-0
© 2022 Elsevier Inc.
All rights reserved.

protection to ensure the device's safety. More recently, an integrated multistage force amplification frame was developed (Wen and Xu, 2018) to achieve a high safety factor and large force amplification ratio along with mechanism protection. However, because the components are connected, the device experiences excessive contact that may eventually damage the components, causing maintenance problems. Moreover, relative motion between the wedge mechanisms restricts the amplification effect and creates noises that affect the walking comfort of the user. Hence, it is desirable to improve the devices with a high level of stability and reliability.

In practice, walking force acting on the floor is imposed by people's weight (in a vertical direction) as well as the relative friction force (in a horizontal direction) between the shoes and the floor. Thus, a higher output power will be achieved if the device can scavenge energy from multiple directions. Nevertheless, most available piezoelectric energy harvesters can harvest vibration energy only from a direct foot strike (in the vertical direction), which results in a loss of harvestable energy. Therefore, it is necessary to design a 2-D piezo stack energy harvester to achieve higher efficiency. Several two-degree-of-freedom (2-DOF) piezo energy harvesters were reported using unidirectional excitation (Hu et al., 2017; Wang and Tang, 2017; Xiao et al., 2015). The major challenge is harvesting energy from excitations in two directions leads to complexity in the energy harvesting device because of the restriction of one pair of input and output. Although a 2-D energy harvester can easily be realized by mounting two 1-D piezo harvesters orthogonally (Su and Zu, 2014), only one piezo stack works to harvest energy in each direction, which results in a relatively low power density.

This chapter presents a new 2-D piezoelectric energy harvester using a single piezoelectric stack to harvest mechanical energy from human footsteps during walking. To maintain the magnitude of the applied force no matter how the walking force changes, permanent magnets are used to produce the attractive force that acts as the input excitation to the piezo stack. By taking advantage of the magnetic input, the safety of the harvester device is guaranteed. To magnify the input force to achieve a higher output power, a planer type of multistage force amplification frame is presented. It is constructed by combining a leverage mechanism and bridge-type force amplifier. In addition, a guiding mechanism is introduced to the frame to prevent forces in other than the input direction. An optimal design for the main parameters for the harvester device is carried out to maximize the force amplification ratio under other constraints.

9.2 Mechanical design

The schematic diagram of proposed 2-D stack-based energy harvester is shown in Fig. 9.1. It contains a multistage force amplification frame with a linear guiding mechanism, four permanent magnets, a top plate, and a bottom plate. The frame is fixed on the bottom plate. Two permanent magnets are mounted on both input ends. Two other magnets are mounted on the top plate.

9.2.1 Working principle

The working principle of the presented piezo stack-based energy harvester is explained as follows. At the initial stage, the magnetic input is maximized owing to the shortest distance between the two adjacent magnets with

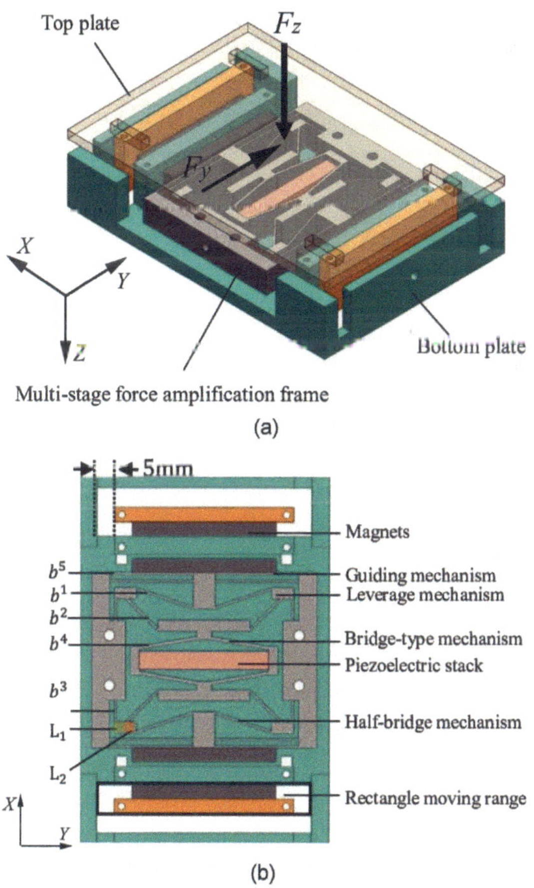

Figure 9.1 Schematic diagram of designed two-directional piezo stack-based energy harvester. (A) Isometric view; (B) Front view.

reverse poles. It results in separation of the top and bottom plates, which are kept at a certain distance to meet the design requirements. Besides the attraction force, other forces between the magnets will not lead to movement of the top plate, which ensures the stability of the device.

When an external force (F_z or F_y) (Fig. 9.1A) with a magnitude larger than the attraction is applied to the top plate, the top plate can move freely within a small rectangular range in the XY plane (Fig. 9.1B). This ensures the walking comfort of the user, because a large change in displacement is not allowed. Moreover, because this movement leads to a change in the distance between magnets, the magnetic force is decreased. The bottom corner of the rectangular range indicates the maximum displacement change, which corresponds to the minimum value of the magnetic force. Once the external force (F_z or F_y) is removed from the harvester, the top plate will immediately be restored to its original position owing to the magnetic attraction force. During one complete press-and-release cycle with the magnetic force acting on the force amplification frame (in the X direction), electrical power is generated by the piezoelectric stack based on the forward piezoelectric effect.

9.2.2 Mechanism design

The mechanical design of each submechanism is presented in detail next.

9.2.2.1 Permanent magnets

The core objective of the design is to guarantee the desired movement of the magnets. When the magnets are constrained by a certain distance, the magnets can undergo only a planner motion. Whether the attraction force is lost depends on the planar movement of the magnets. On the contrary, when a magnet approaches the other one (with a reverse pole), the attraction will increase dramatically. In practice, it is hard to control and maintain the distance between magnets.

In this work, the energy harvester is designed by considering 2-DOF motion in the YZ plane. Owing to the characteristics of magnetic attraction, the behavior of the two magnets functions like a spring. Once the external force is released, the magnets will return to their original positions as long as displacement between the magnets does not exceed the limit value. In this case, the external force (F_z or F_y) is more like a switch to control the position of the top plate, resulting in the displacement change between magnets. More important, owing to the noncontact nature between magnets, movement of the top plate is smoother than that of the contact design in the previous work. This also ensures the walking comfort of people.

9.2.2.2 Multistage force amplification frame with linear guiding mechanism

The adopted multistage force amplification frame contains three parts: a half-bridge mechanism, a leverage mechanism, and a bridge-type force amplifier. It is introduced to achieve a large force amplification ratio with an allowable safety factor. Because of the large attraction force between magnets, a linear guiding mechanism is used to reduce the influence of the force and prevent the piezoelectric stack from shear forces. In particular, the linear guiding mechanism is constructed of two groups of four leaf flexures, which are connected to both input ends of the force amplification frame.

The force amplification frame can be regarded as an improved version of those in the literature (Wen and Xu, 2018; Xu, 2017). Each leaf flexure acts like a fixed-guided flexure beam. The linear guiding mechanism can produce purely translational motion, because the stiffness of each leaf spring in the moving direction is much smaller than stiffness in other directions.

9.2.2.3 Protection mechanism

With a protection mechanism, the proposed energy harvester is better than those in previous work in terms of device safety. Fig. 9.2A shows that no other components (such as screws) block the motion of the top plate. When the top plate is fully compressed, there is more bottom area support available to withstand the large stress. Because human weight is not applied directly to the force amplification frame, the piezoelectric stack and force amplification frame are well-protected for any special situation. Owing to the attraction force between magnets, the device is feasible for harvesting energy from excitation with 2-DOF motion. Furthermore, to achieve a compact size, the rectangle moving range in the bottom plate is hollow, as shown in Fig. 9.2B. Hence, the width of the bottom layer can be ignored; it can reduce the height of the structure. Gaps between the top plate and the frame are adopted to ensure normal working motion and force amplification performance.

Overall, most of the protection functions are realized by the bottom plate. The bottom plate is used to fix the multistage force amplification frame, restrict the movement of the top plate, hold a constant distance between the magnets, and bear a large stress. In addition, the top plate serves as an interface to apply the human weight and experiences movement in the YZ plane.

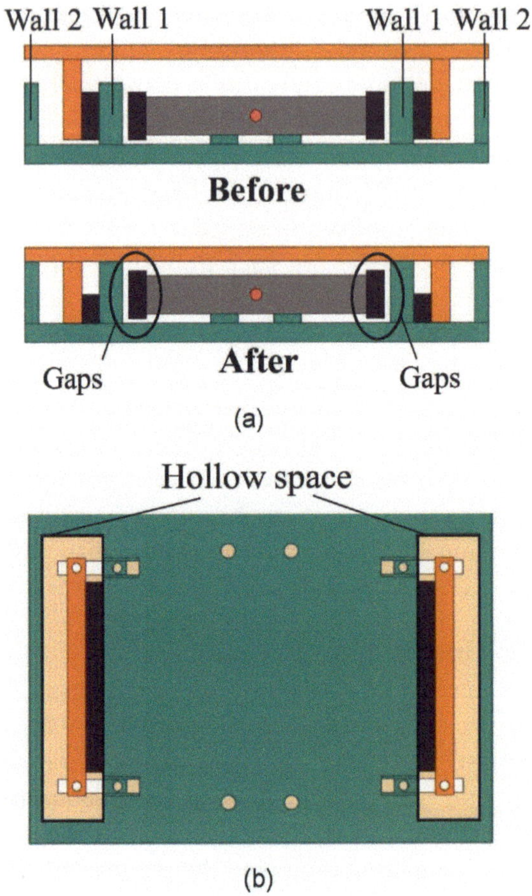

Figure 9.2 Design details of the energy harvester. (A) Side view of device before and after exerted force; (B) Hollow space in the bottom plate.

9.2.3 Material selection

Owing to the multicomponent feature of the energy harvester, various types of materials can be selected. Because of the relatively light weight and high tensile yield strength, aluminum 7075 was selected to fabricate the multistage force amplification frame. For rapid prototyping, acrylonitrile butadiene styrene plus was adopted for fabricating the top and bottom plates through three-dimensional printing. A piezoelectric stack (RP150/5x5/30, Harbin Soluble Core Technology Co., Ltd.) was selected owing to its relatively large value of d33 coefficient. The main properties of the adopted piezoelectric stack are listed in Table 9.1.

Table 9.1 Main properties of piezoelectric stack in d_{33} mode.

Property	Symbol	Value
Relative permittivity	$\varepsilon_{33}^{T}/\varepsilon_0$	3500
Coupling factor	k_{33}	0.7
Capacitance	C_p	2.4×10^{-6} F
Elastic compliance coefficient	S_{33}^{E}	18.5×10^{-12} m^2/N
Piezoelectric voltage coefficient	g_{33}	17×10^{-3} Vm/N
Piezoelectric charge coefficient	d_{33}	650×10^{-12} C/N

9.3 Analytical modeling

In this section, the analytical static models of the permanent magnets and multistage force amplification frame are described in detail.

9.3.1 Model of magnetic attraction

A schematic diagram of a pair of magnets is shown in Fig. 9.3. According to Stanton et al. (2010) and Yung et al. (1998), the mathematical model of the permanent magnets can be established as:

$$M = B_r/\mu_0 \tag{9.1}$$

$$\mu = MV \tag{9.2}$$

where B_r denotes the residual flux density of the magnet; μ is the magnetic moment vector; μ_0 represents the permeability of the clearance; M denotes

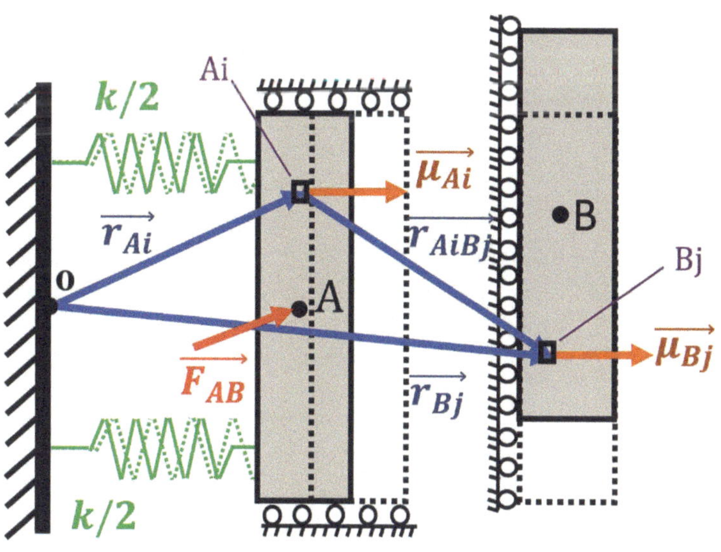

Figure 9.3 Illustration of static model for magnetic attraction.

the magnetization vector, which is the sum of the microscopic magnetic moment vector of the magnet's mesh; and V describes the volume of the magnet's mesh.

In Fig. 9.3, assume that magnet A moves horizontally and is connected to a spring that represents the force amplification frame. Moreover, magnet B maintains a distance relative to magnet A and moves vertically. Then, position vectors $\vec{r_{Ai}}$, $\vec{r_{Bj}}$, and $\vec{r_{AiBj}}$ can be derived as:

$$x_A = F_x/k \tag{9.3}$$

$$\vec{r_{Ai}} = [x_A\, y_{Ai}\, z_{Ai}] \tag{9.4}$$

$$r_{Bj} = \left[d\ y_{Bj} + y_B\ z_{Bi}\right] \tag{9.5}$$

$$\vec{r_{AiBj}} = \vec{r_{Bj}} - \vec{r_{Ai}} \tag{9.6}$$

where F_x denotes the X-axis component of attraction between the two magnets, x_A is the deflection of the spring, k is the stiffness of the frame, d represents displacement between the magnets without the impact of magnetic attraction, and y_B denotes the vertical movement of magnet B.

Because the mesh volumes of magnets A and B are the same, magnetic dipole moment vectors μ_{Ai} and μ_{Bj} can be obtained as:

$$\vec{\mu_{Ai}} = \vec{\mu_{Bj}} = [\mu\ 0\ 0] \tag{9.7}$$

At the position of magnet B, the magnetic field produced by magnet A can be written as:

$$\vec{B_{AiBj}} = -\frac{\mu_0}{4\pi}\nabla\frac{\vec{\mu_{Ai}}\cdot\vec{r_{AiBj}}}{\left|\vec{r_{AiBj}}\right|_2^3} \tag{9.8}$$

where $||\cdot||_2$ represents the Euclidean norm, and ∇ denotes the vector gradient.

In addition, the magnetic potential energy ($U_{m,\ AiBj}$) of magnets A and B and the attraction force (F_{AB}) of the two magnets can be expressed as:

$$U_{m,AiBj} = -\vec{B_{AiBj}}\cdot\vec{\mu_{Bj}} \tag{9.9}$$

$$F_{AB} = [F_x\ F_y\ F_z] = \sum_{Ai=1}^{n}\sum_{Bj=1}^{n}\frac{\partial U_{m,AiBj}}{\partial r}\cdot\frac{\vec{r_{AiBj}}}{\left|\vec{r_{AiBj}}\right|} \tag{9.10}$$

9.3.2 Modeling of multistage force amplification frame

In previous work (Wen and Xu, 2018), an analytical model of the multistage force amplification frame was established by the stiffness matrix method. The model is used to describe the relationship between the force and displacement of each flexure component and to determine the parameters for further optimization. Every flexure component is regarded as the fundamental frame element connected by two nodes. A schematic diagram of the frame is shown in Fig. 9.4. In to Fig. 9.5, force vector $\{\mathbf{F}^i\}$ and displacement vector $\{\mathbf{U}^i\}$ for node i are given by:

$$\{F^i\} = \left[F_x^i \; F_y^i \; M_z^i\right]^T \tag{9.11}$$

$$\{U^i\} = \left[\; U_x^i \; U_y^i \; \theta_z^i \;\right]^T \tag{9.12}$$

with $i = 1, 2, 3 \ldots, n$, where n represents the number of nodes.

Furthermore, force vector **F** and displacement vector **U** for all nodes can be expressed as:

$$U = [U^1 \ldots U^i \ldots U^n]^T \tag{9.13}$$

$$F = [F^1 \ldots F^i \ldots F^n]^T \tag{9.14}$$

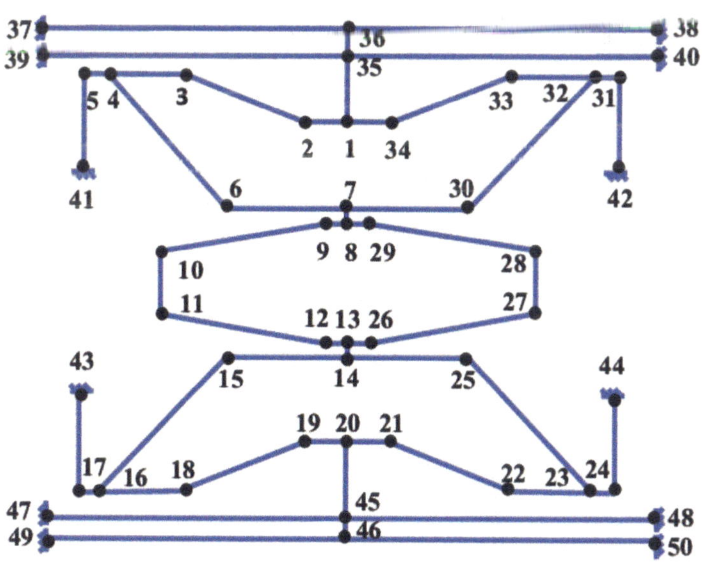

Figure 9.4 Schematic of multistage force amplification frame.

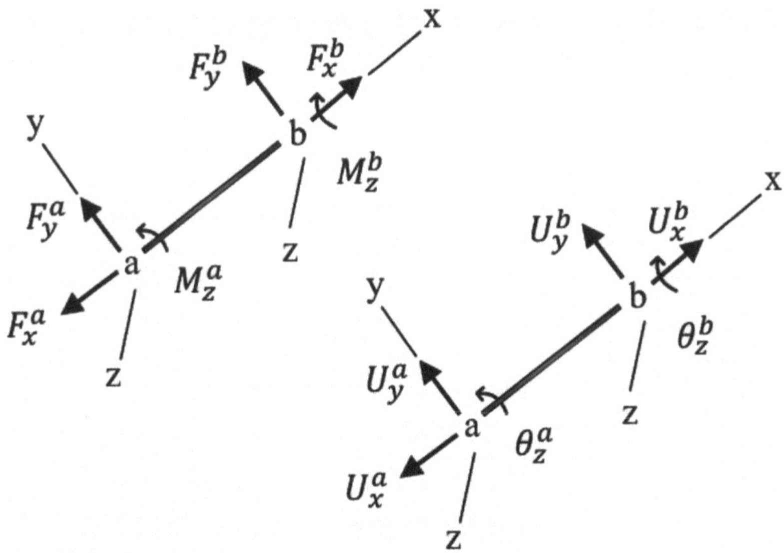

Figure 9.5 Frame element with two related nodes.

The global system equation of the frame can be written as:

$$F = KU \tag{9.15}$$

where global stiffness **K** can be derived based on the stiffness matrix method (Wen and Xu, 2018).

The boundary conditions are set considering the input force and fixed support. Finally, the amplification ratio of leverage n_{leverage}, bridge-type force amplifier n_{bridge}, and the entire force amplification frame, n_{frame} can be calculated as:

$$n_{\text{leverage}} = \frac{F^4}{F^1} \tag{9.16}$$

$$n_{\text{bridge}} = \frac{F^{10}}{F^8} \tag{9.17}$$

$$n_{\text{frame}} = \frac{F^{10}}{F^1} \tag{9.18}$$

9.4 Parameter optimization and finite element analysis simulation study

After the structural parameters are derived from analytical models, the design of the parameters is optimized in this section. Moreover, an FEA simulation study of the optimal energy harvester is carried out and compared with previous work.

9.4.1 Dimension optimization

9.4.1.1 Permanent magnets

During the operation, a large displacement difference between the two magnets results in a large force output and power output. In contrast, a small displacement difference leads to more comfort when the user walks. Thus, a trade-off is needed to assign the input stroke for the energy harvester. In this work, the input strokes of the harvester in both the Y and Z directions are assigned to be 5 mm. Because of the stiffness of the frame and the attraction force, displacement between the magnets is derived as 9.2 mm.

Based on the parameters of the permanent magnets listed in Table 9.2, the maximum values of $|F_x|$, $|F_y|$, and $|F_z|$ are obtained with MATLAB® software. The maximum force indicates the capability of the device. In addition, the largest difference value of the maximum force can be calculated, which represents the input force applied to the force amplification frame, because only the force difference leads to electrical power generation for the piezoelectric stack. During work, the applied shear forces in the Y and Z directions are transmitted into the X-direction magnetic force, which is exerted onto the piezoelectric stack by the force amplification frame and contributes to power generation.

The states of the top plate at typical phases are depicted in Fig. 9.6. The attraction forces for various values of input displacement y_B and directions are shown in Fig. 9.7A. The maximum value of 20.82 N can be achieved for $|F_x|$ when the magnets are located at the central point, as shown in

Table 9.2 Main parameters of adopted permanent magnets.

Parameter	Symbol	Value
Displacement between magnets	d	9.2 mm
Frame's stiffness	k	83.34 N/mm
Volume	V	1157.62 mm^3
Residual flux density	B_r	1.22 T
Permeability of clearance	μ_0	1.256×10^{-6} H/m

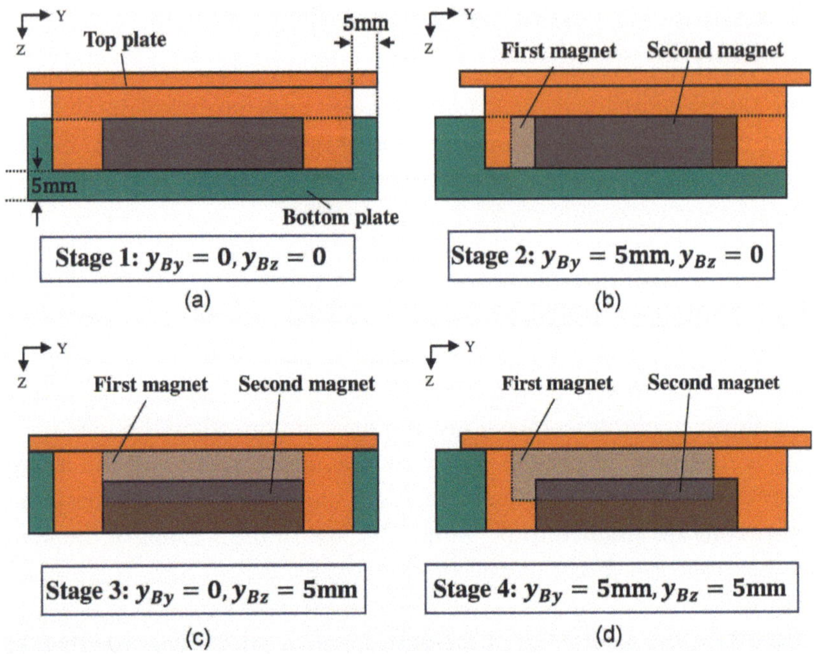

Figure 9.6 Four typical phases of movement for the top plate. (A) Phase 1; (B) Phase 2; (C) Phase 3; (D) Phase 4.

Fig. 9.6A. For movement of the top plate along the Y direction, as depicted in Fig. 9.6B, when the magnet is moved from the central point, $|Fx|$ is generally decreased until a maximum displacement of 5 mm is reached, in which the minimum value of 18.23 N is obtained for $|Fx|$. Thus, the largest difference value of the maximum force for $|Fx|$ is 2.59 N, which is the real force exerted on the force amplification frame.

Similarly, as the top platform translates along the Z direction, as illustrated in Fig. 9.6C, the induced minimum attraction force for $|Fx|$ is 15.57 N, as shown in Fig. 9.7B. The largest difference value of the maximum force of $|Fx|$ is 5.25 N, which contributes to piezoelectric power generation.

In addition, the maximum shear force is reached as the top platform translates along the Z direction, where $|F_z|$ is 5.08 N. Thus, to reduce the influence of the shear force (F_y or F_z), a guiding mechanism is required.

9.4.1.2 Multistage force amplification frame

The multistage force amplification frame contains multiple flexure beams that seriously influence the force amplification ratio. According to the

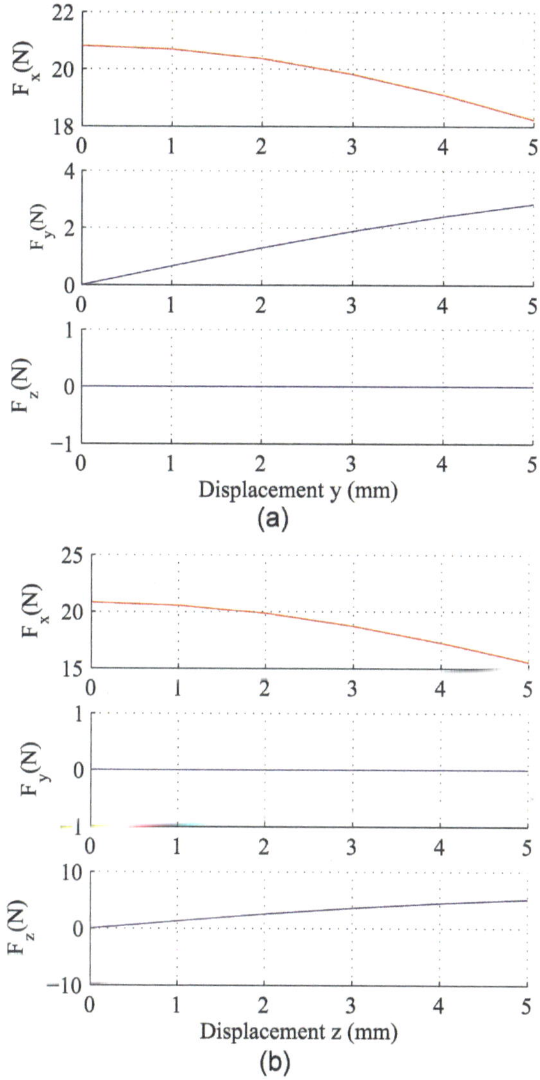

Figure 9.7 Attraction forces with various input displacements and directions. (A) Along the Y direction; (B) Along the Z direction.

analytical model, the width (w), length (l), height (h), and thickness (t) of the thin flexure beams are selected as the key parameters. Concerning the leverage mechanism, input arm L_1 and output arm L_2 of the simplified flexure beams are also important. Furthermore, to simplify optimization, the height is set to 8 mm to ensure a large safety factor for the material.

Finally, a total of 14 structural parameters are selected. According to (Wen and Xu, 2018), the multiobjective genetic algorithm is implemented for multiparameter optimization because it is suitable for multiconstraint optimization problems. Moreover, the primary objective is to maximize the force amplification ratio under several constraints so that the device can be more stable and produce more power. The corresponding optimization procedure is thus:

- Objective: To maximize force output
- Parameters: 14 structural parameters of flexure beams
- Subject to:
  1. Force = 100 N;
  2. Safety factor >2.

Based on the assigned objective and constraints, parameter optimization starts with 1000 samples to find three candidates in the reasonable range. The initial value, range, and optimal results of the parameters are shown in Table 9.3.

9.4.2 Finite element analysis simulation results

In practice, external forces applied on the top platform of the harvester along the Y and Z directions are transmitted into X-directional magnetic force, which contributes to power generation. Thus, in an FEA simulation study, the output performance of the device is examined with an input

Table 9.3 Parameter optimization results for the energy harvester.

Parameter	Range	Initial value	Optimal result
Length of b^1 (mm)	[5:0.1:16]	12	16
Width of b^1 (mm)	[4:0.1:6]	5	5.2
Thickness of b^1 (mm)	[0.5:0.1:1]	0.5	0.5
Length of b^2 (mm)	[7:0.1:12]	9	9.2
Width of b^2 (mm)	[5:0.1:12]	8	8.8
Thickness of b^2 (mm)	[0.5:0.1:1]	0.5	0.5
Length of b^3 (mm)	[4:0.1:6]	5	6
Thickness of b^3 (mm)	[0.5:0.1:1]	0.5	0.5
Length of b^4 (mm)	[14:0.1:16]	15	16
Width of b^4 (mm)	[2.5:0.1:3]	2.5	2.7
Thickness of b^4 (mm)	[0.5:0.1:1]	0.5	0.8
Input arm length L_1 (mm)	[2:0.1:6]	3	2.6
Output arm length L_2 (mm)	[2:0.1:6]	3	2.2
Thickness of b^5 (mm)	[0.5:0.1:1]	0.5	0.5

force applied in the X direction. In addition, input forces in the Y and Z directions are considered to be shear forces that are along directions other than those that move.

In particular, the experienced stress of the multistage force amplification frame is tested under the maximum attraction force, as shown in Fig. 9.8.

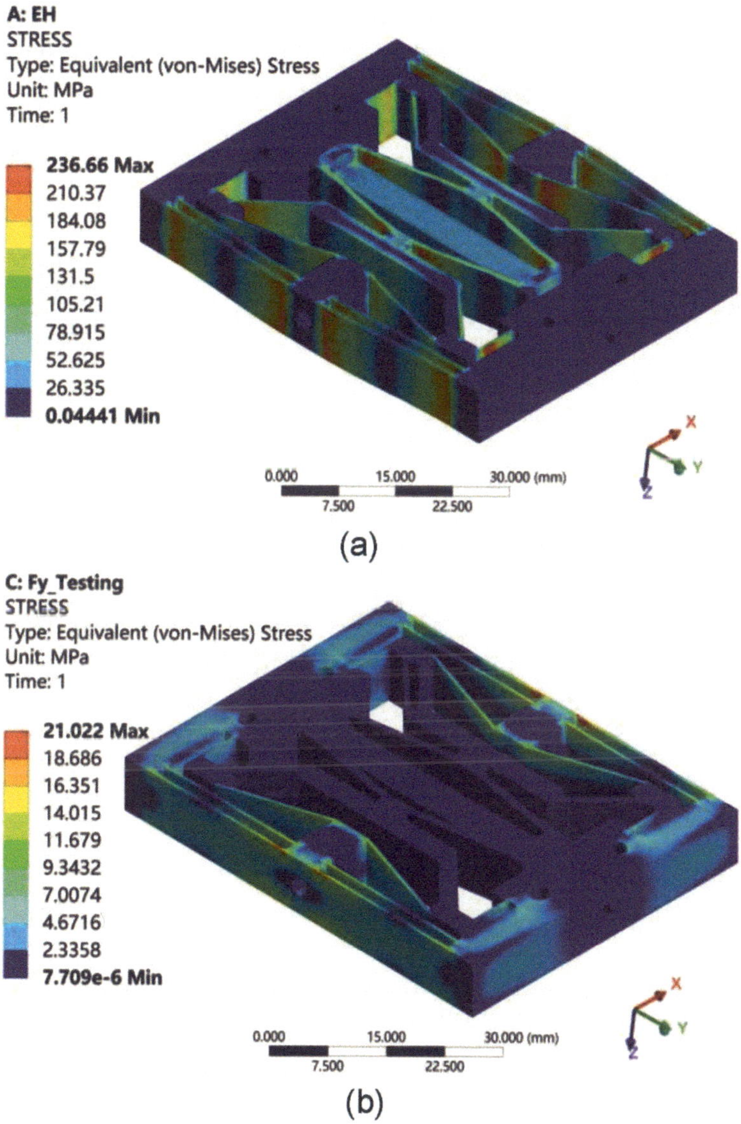

Figure 9.8 Simulation results of stress distribution for the multistage force amplification frame. (A) The frame is applied by $|F_x| = 100$ N; (B) The frame is applied by $|F_z| = 30$ N.

First, an input force of $F_x = 100$ N is applied to both input ends of the frame and the pivots of the leverage are fixed. The result shows that the maximum stress is 236.66 MPa. This indicates a safety factor of 2.12, which is sufficient for long-term use. Moreover, simulation results show that the multistage force amplification frame provides a force amplification ratio of 10.54.

Second, in view of the influence of the force (in the Y and Z directions) in other than moving directions, a shear force of 30 N along the Z direction is applied to test the output performance of the harvester. Simulation results of the stress distribution for the frame are depicted in Fig. 9.8B. The maximum stress of 22 MPa further demonstrates that the frame is able to bear a large shear force. In addition, the output performance is not affected owing to the introduction of the linear guiding mechanism.

9.5 Prototype fabrication and experimental study

9.5.1 Prototype fabrication

A prototype of the designed 2-D energy harvester was fabricated; the experimental setup is illustrated in Fig. 9.9. For performance testing, the harvester was driven by a voice coil motor (NCC05-18-060-2X, H2W Techniques, Inc.), which provided periodic motion with the given frequency. The exerted force on the top plate was measured by a force sensor (LSB200, FUTEK Advanced Sensor Technology, Inc.). The voltage output from the piezoelectric stack across a variable resistance box (ZX21A, Shanghai Precision Instruments, Inc.) was acquired by a digital multimeter (PXIe-4082, National Instruments, Inc.). Finally, all data were collected and analyzed through a real-time controller (PXIe-1082, National Instruments, Inc.).

9.5.2 Experimental results

9.5.2.1 Results of Y-direction excitation

First, the output performance of the energy harvester along the Y direction was tested under an excitation frequency of 5 Hz, as shown in Fig. 9.10. Considering the piezoelectric stack as pure capacitance, optimal resistance Z_p was obtained according to the equation:

$$Z_p = \frac{1}{\omega_p C_p} \tag{9.19}$$

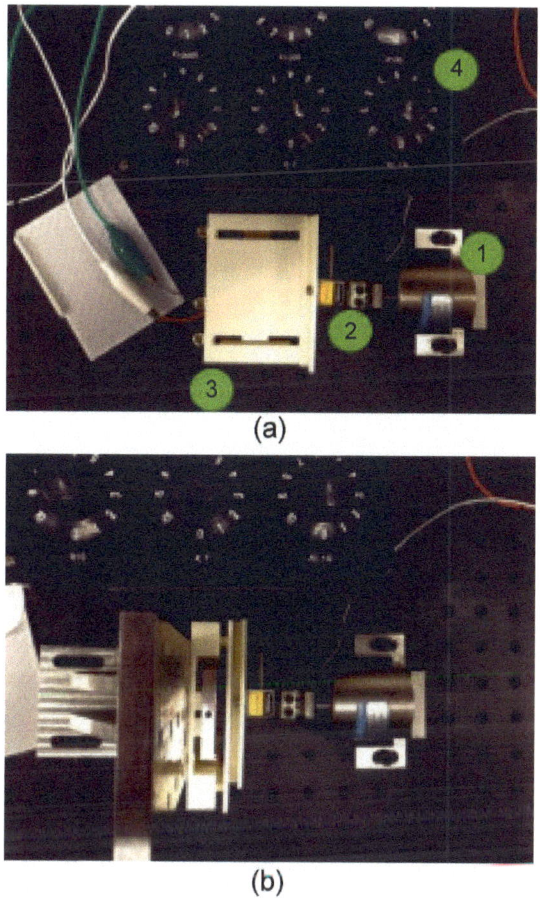

Figure 9.9 Fabricated prototype and experimental setup of energy harvesting device. (A) Experimental setup for testing output performance under Y-directional excitation with (1) voice coil actuator, (2) force sensor, (3) prototype of fabricated energy harvester, and (4) resistance box; (B) Experimental setup for testing output performance under Z-directional excitation.

where C_p denotes the piezoelectric capacitance and ω_p represents the excitation frequency in units of rad/s.

Under an excitation frequency of 5 Hz and piezoelectric capacitance of 2.4 μF, the calculated optimal resistance was 13.26 kΩ. The maximum power was achieved under optimal resistance according to the resistance matching theory. In the experiment, a maximum peak power output of 2.49 μW and maximum average power of 0.27 μW were measured across a resistance of 13 kΩ, which is close to the calculated optimal value.

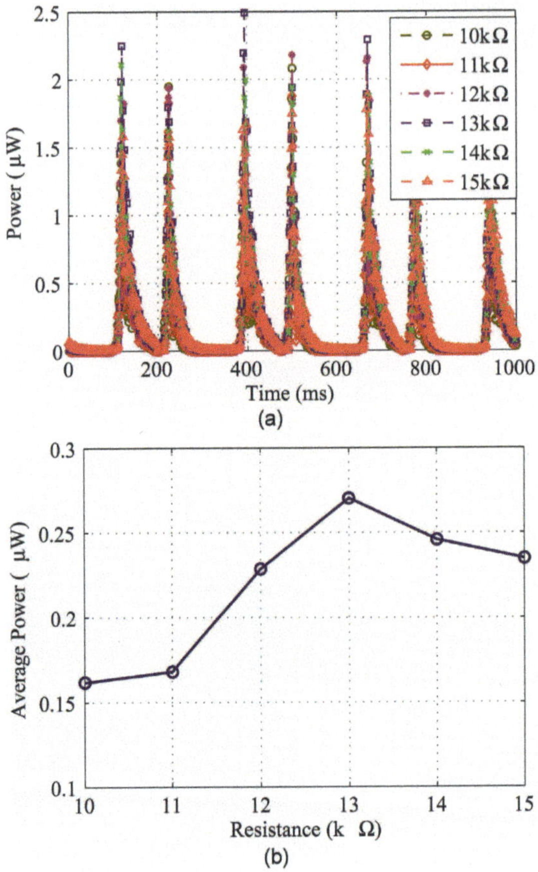

Figure 9.10 Output performance of energy harvester with Y-directional excitation. (A) Time-history results; (B) Average power versus loading resistance.

However, the captured energy was limited because the adopted rectangle type of magnets led to a smaller change in attraction force when the input displacement of 5 mm was experienced. On the other hand, magnets required more recovery time when the change in attraction force was small. Under an excitation with high frequency, the energy harvesting capacity was reduced because the top plate could not return to its original position before the start of the next cycle. The shortened displacement between magnets resulted in less harvested power. Therefore, the harvester is more suitable for low-frequency application (e.g., smaller than 10 Hz).

In addition, for the entire structure, the wall (Fig. 9.2A) was required to keep the displacement of the magnets. Although the wall–magnet connection did not affect the power output directly when an input force was exerted, friction led to relative motion and restricted the covering time, which resulted in the loss of generated energy.

9.5.2.2 Results of Z-direction excitation

Similarly, the output performance of the energy harvester along the Z direction was tested; the results are shown in Fig. 9.11. The maximum peak power output of 8.47 μW and the maximum average power of 2.66 μW were obtained across a loading resistance of 13 kΩ. Compared with the

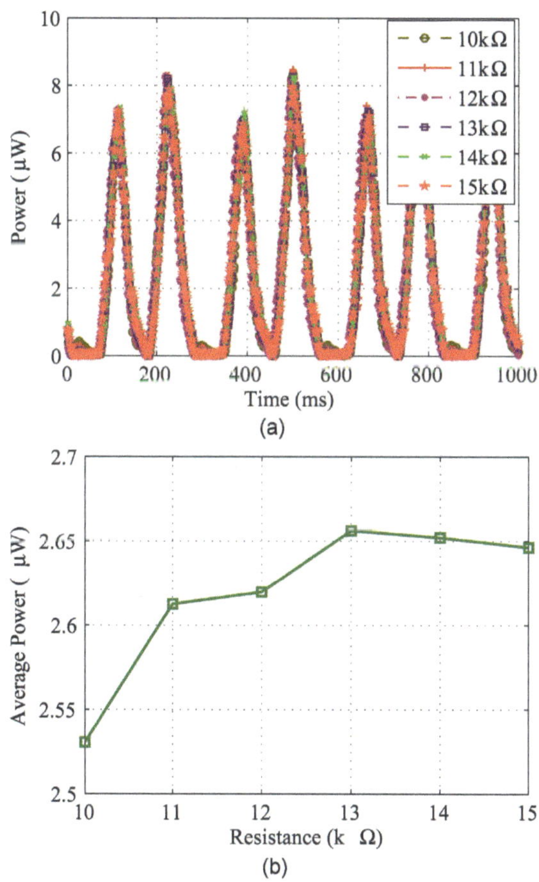

Figure 9.11 Output performance of energy harvester with Z-directional excitation. (A) Time-history results; (B) Average power versus loading resistance.

results of output performance in the Y direction, although friction still existed, which could influence the entire output performance, the energy harvester generated more energy output owing to a larger difference in attraction force. In addition, the maximum power occurred when the loading resistance was close to the calculated optimal value of resistance.

9.5.3 Further discussion

Because of the highly flexible movement of the top plate, the device harvested energy from excitations of 2-DOF motion with one piezoelectric stack. No matter how large an impact force was exerted on the harvester, the input force to the harvester was provided by the difference in attraction force between magnets. Such a feature further protects the energy harvester in practical applications. Yet, in this design, only one pair of magnets is used and tested. Thus, the generated power is relatively small (around 10 μW). To overcome this issue, it is desirable to increase the numbers and residual flux density of the magnets. Moreover, it is suggested to improve the attraction difference by adjusting displacement between the magnets. This can reduce the recovery time and enhance the input force, leading to larger output power.

The experimental study shows that an auxiliary mechanism that assists the resilience of the top plate can achieve a fast recovery speed and result in more energy harvesting. Because the linear guiding mechanism is required to eliminate the impact of force in other than moving directions, the output performance is affected owing to the large stiffness of the flexure mechanism. A possible approach is to use the compound parallelogram flexure mechanism, which can reduce its own stiffness and maintain a large linear motion. In future work, such an improved design will be carried out by considering both the output performance and the compactness of the device.

9.6 Conclusion

This chapter presents the mechanical design, parameter optimization, prototype fabrication, and experimental study of a 2-D piezoelectric energy harvester with permanent magnets and a multistage force amplification frame. An FEA simulation study and experimental study demonstrate the output performance of the energy harvester. Moreover, the proposed device has smooth conversion from the excitations of both directions. Under the restriction of input force and input displacement, the reliability and feasibility of the energy harvester are further improved compared with previous designs. In future work, it is desirable to improve the power output by increasing the magnetic force of the magnets. The reported design methodology can also be extended to the design of 3-DOF energy harvesters.

References

Chen, W., Wang, Y., & Deng, W. (2017). Deformable force amplification frame promoting piezoelectric stack energy harvesting: Parametric model, experiments and energy analysis. *Journal of Intelligent Material Systems and Structures, 28*(7), 827–836. https://doi.org/10.1177/1045389X16657426

Feenstra, J., Granstrom, J., & Sodano, H. (2008). Energy harvesting through a backpack employing a mechanically amplified piezoelectric stack. *Mechanical Systems and Signal Processing, 22*(3), 721–734. https://doi.org/10.1016/j.ymssp.2007.09.015

Hu, G., Tang, L., Das, R., & Marzocca, P. (2017). A two-degree-of-freedom piezoelectric energy harvester with stoppers for achieving enhanced performance. *International Journal of Mechanical Sciences, 149*, 500–507.

Stanton, S. C., McGehee, C. C., & Mann, B. P. (2010). Nonlinear dynamics for broadband energy harvesting: Investigation of a bistable piezoelectric inertial generator. *Physica D: Nonlinear Phenomena, 239*(10), 640–653. https://doi.org/10.1016/j.physd.2010.01.019

Su, W. J., & Zu, J. W. (2014). Design and development of a novel bi-directional piezoelectric energy harvester. *Smart Materials and Structures, 23*(9). https://doi.org/10.1088/0964-1726/23/9/095012

Wang, H., & Tang, L. (2017). Modeling and experiment of bistable two-degree-of-freedom energy harvester with magnetic coupling. *Mechanical Systems and Signal Processing, 86*, 29–39. https://doi.org/10.1016/j.ymssp.2016.10.001

Wang, X., Shi, Z., Wang, J., & Xiang, H. (2016). A stack-based flex-compressive piezoelectric energy harvesting cell for large quasi-static loads. *Smart Materials and Structures, 25*(5). https://doi.org/10.1088/0964-1726/25/5/055005

Wang, L., Chen, S., Zhou, W., Xu, T. B., & Zuo, L. (2017). Piezoelectric vibration energy harvester with two-stage force amplification. *Journal of Intelligent Material Systems and Structures, 28*(9), 1175–1187. https://doi.org/10.1177/1045389X16667551

Wen, S., & Xu, Q. (2018). Design of a novel piezoelectric energy harvester based on integrated multi-stage force amplification frame. *IEEE/ASME Transactions on Mechatronics, 24*(3), 1228–1237. https://doi.org/10.1109/TMECH.2019.2906999

Wen, S., Xu, Q., & Zi, B. (2018). Design of a new piezoelectric energy harvester based on compound two-stage force amplification frame. *IEEE Sensors Journal, 18*(10), 3989–4000. https://doi.org/10.1109/JSEN.2018.2820221

Xiao, H., Wang, X., & John, S. (2015). A dimensionless analysis of a 2DOF piezoelectric vibration energy harvester. *Mechanical Systems and Signal Processing, 58*, 355–375. https://doi.org/10.1016/j.ymssp.2014.12.008

Xu, Q. (2017). Design and development of a novel compliant gripper with integrated position and grasping/interaction force sensing. *IEEE Transactions on Automation Science and Engineering, 14*(3), 1415–1428. https://doi.org/10.1109/TASE.2015.2469108

Yung, K. W., Landecker, P. B., & Villani, D. D. (1998). An analytic solution for the force between two magnetic dipoles. *Magnetic and Electrical Separation*, 39–52. https://doi.org/10.1155/1998/79537

CHAPTER 10

Design of a dual-axial underfloor piezoelectric energy harvester

This chapter presents the design, fabrication, and testing of a dual-axial underfloor piezoelectric energy harvester based on a piezoelectric stack. Usually, an underfloor energy harvester can scavenge the energy only from vertical force excitation. In addition to vertical pressure, the floor experiences a lateral friction force during human walking. The underfloor energy harvester can collect energy from both vertical and horizontal excitation forces. To amplify the input force to obtain larger output power, a flexure-based two-stage force amplifier (TSFA) is introduced and optimally designed. A prototype was fabricated and tested. Experimental results demonstrate the promising performance of the developed energy harvester.

10.1 Introduction

To harvest energy from the footsteps of a human walking, a piezoelectric stack-based piezoelectric energy harvester has the advantage of small deflection. It is preferable because a small vertical deflection will not much influence the way people walk. Because the frequency of human walking is low (less than 10 Hz), piezoelectric stack-based piezoelectric energy harvesters are usually excited directly by an external force (Wen et al., 2018; Wu and Xu, 2019) rather than ambient vibration in conventional vibration-based energy harvesters (Yang and Zu, 2014). Most available piezoelectric energy harvesters enable energy harvesting only from uniaxial force excitations (Chen et al., 2017; Wang et al., 2017; Wen and Xu, 2019). During human walking on a floor, foot strikes exert a vertical force (owing to weight) and also a lateral friction force on the floor. Hence, it is best to devise a multiaxial piezoelectric energy harvester (Saadatzi et al., 2018) to scavenge energy from more excitation directions of input force. Because the use of multiple piezoelectric stacks to achieve a multiaxial piezoelectric energy harvester is not economical, some dual-axial and

Mechanical Design of Piezoelectric Energy Harvesters
ISBN 978-0-12-823364-1
https://doi.org/10.1016/B978-0-12-823364-1.00009-2
© 2022 Elsevier Inc.
All rights reserved.

triaxial piezoelectric energy harvesters were developed using a single piezoelectric stack (Wu and Xu, 2020). The previous piezoelectric energy harvester design (Wu and Xu, 2020) had some shortcomings. For instance, the triaxial piezoelectric energy harvester had a large model-prediction error and instability. In addition, considering that energy harvesting from a two-directional (2-D) force or pressure excitation is sufficient, because most roads are straight, a dual-axial piezoelectric energy harvester with greater output power is desirable for practical application.

This chapter reports a two-axial underfloor energy harvester that can scavenge energy from both footstep force (in the vertical direction) and friction force (in the horizontal direction) imposed by people walking. Specifically, the underfloor energy harvester is composed of several conical transmission mechanisms (CTMs), two TSFAs, and two piezoelectric stacks. Unlike previous work, the sliding guiding mechanisms are replaced with slider mechanisms to reduce friction inside the harvester mechanism. The three-beam frame with a nonnotch type of flexure hinge is adopted in the TSFA to meet the expected physical size demand (Wen et al., 2018). Position limiters are introduced to constrain the maximum input displacement/force of the input plate to protect the designed mechanism. Moreover, the CTMs and slider mechanisms are used to change and guide the dual-axial directions of the input force. The outcome direction of the CTMs and slider mechanisms is in line with the input direction of the TSFAs. Two-stage force amplifiers are adopted to magnify force exerted on the embedded piezoelectric stack. In this way, the piezoelectric stack generates more electric energy. Moreover, the parameters of the TSFA were optimally tuned by a genetic algorithm (GA) under certain constraints. A simulation study with finite element analysis (FEA) was carried out to evaluate the performance of the proposed underfloor energy harvester. A prototype dual-axial underfloor energy harvester was fabricated and its performance was characterized. A practical application test of the underfloor energy harvester was conducted for demonstration.

10.2 Mechanism design

This section introduces the mechanism design and working principle of the proposed underfloor energy harvester. A computer-aided design model of the underfloor energy harvester is depicted in Fig. 10.1.

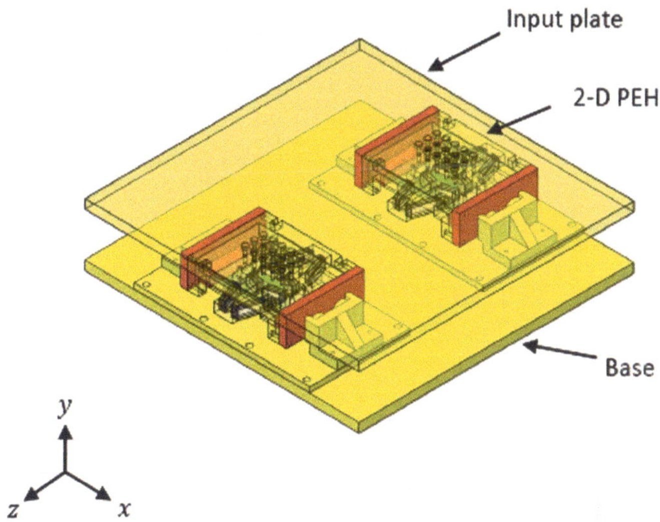

Figure 10.1 Computer-aided design model of proposed underfloor energy harvester.

10.2.1 Working principle

As shown in Fig. 10.1, the proposed underfloor energy harvester consists of two identical dual-axial piezoelectric energy harvesters. For each dual-axial piezoelectric energy harvester, the conical groove ends of the CTMs are connected to the upper input plate and coupled to the position limiters, as illustrated in Fig. 10.2A. The conical ends of the CTMs are also linked to the input ends of a TSFA and the sliders of slider mechanisms. A piezoelectric stack is embedded in the TSFA to generate electric power.

The inside top view of a 2-D piezoelectric energy harvester with a force applied in the negative z-axis direction is illustrated in Fig. 10.2B. A work flowchart of the underfloor energy harvester is shown in Fig. 10.3. When the proposed underfloor energy harvester is excited by an input force with y-axis and z-axis components, the exceeding magnitude of the input displacement (and force) will be constrained by the position limiters, and the rest of the input displacement (and force) will be transmitted into the CTM. Then, a reaction force will be generated on the conical surface of the CTM. Moreover, because the conical ends of the CTMs are connected to the sliders of slider mechanisms, the TSFA will receive only the x-axis component of the reaction force. Through the magnification of the TSFA, force exerted on the piezoelectric stack will generate electric energy based on the direct piezoelectric effect.

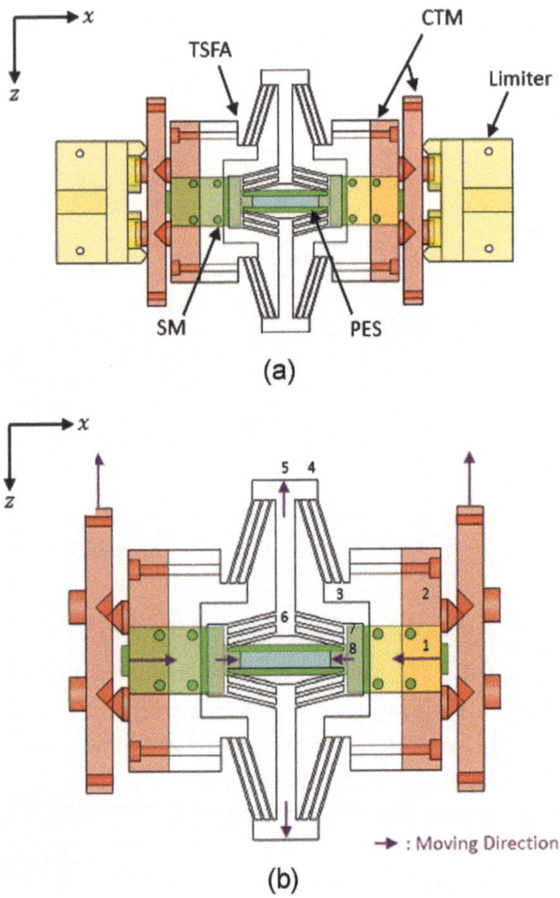

Figure 10.2 Inside top view of a two-directional piezoelectric energy harvester (A) Without force applied and (B) With force applied.

10.2.2 Model of conical transmission mechanism

To facilitate the performance evaluation, an analytical model of the CTM was established. A static model of the CTM is shown in Fig. 10.4, where F_{CTM}, F_{TSFA}, R, and N are the input force of the CTM, input force of the TSFA, reaction force of the CTM, and reaction force of the slider mechanism, respectively. For static equilibrium, the force amplification ratio of the CTM (η_{CTM}) is given as:

$$\eta_{\mathrm{CTM}} = \frac{\cos\beta - \mu_{\mathrm{SM}}\mu_{\mathrm{CTM}}\cos\beta - (\mu_{\mathrm{CTM}} + \mu_{\mathrm{SM}})\sin\beta}{2\sin\beta + 2\mu_{\mathrm{CTM}}\cos\beta} \tag{10.1}$$

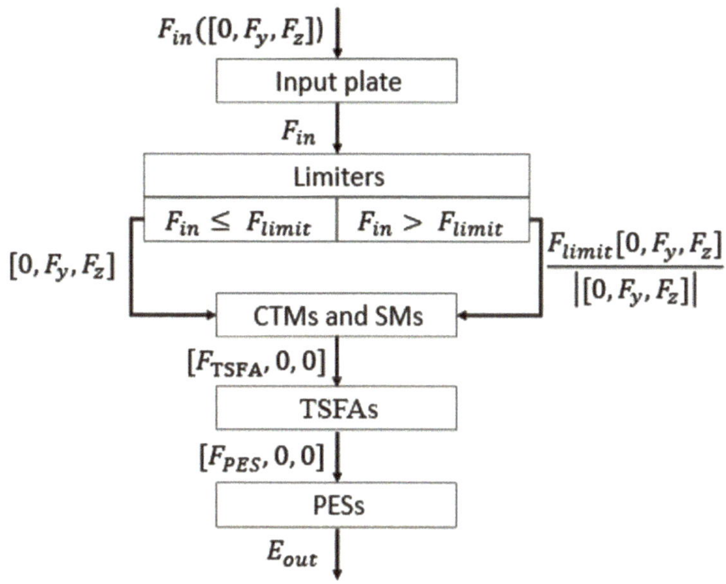

Figure 10.3 Work flowchart of the proposed underfloor energy harvester.

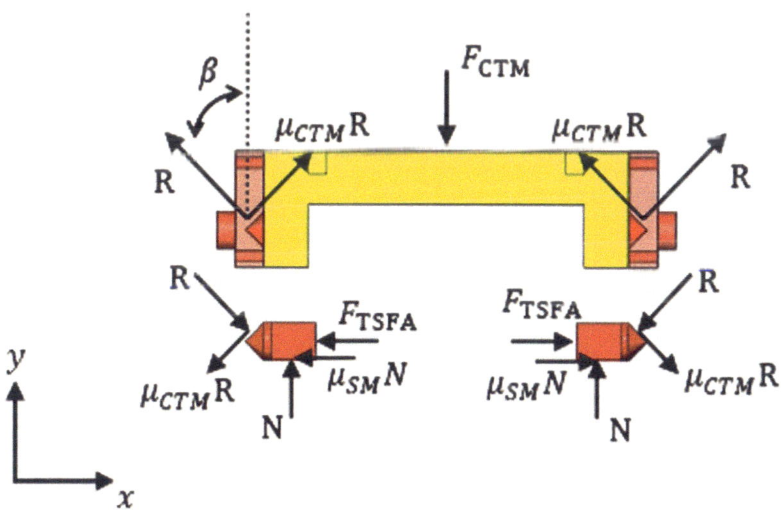

Figure 10.4 Schematic diagram of conical transmission mechanism for static force analysis.

where β is the angle between the y-axis and the inclined plane, and μ_{CTM} and μ_{SM} are the friction coefficients of the sliding surfaces of the CTM and slider mechanism, respectively.

Because the friction coefficient is difficult to predict, the case without a friction coefficient is considered in this study.

10.2.3 Model of two-stage force amplifier

As shown in Fig. 10.2B, the TSFA is symmetric and the hinge parameters of each stage are the same. Hence, the schematic diagram (Fig. 10.5A) of the TSFA is represented by a quarter model with one beam. Then, an analytical model of the TSFA is established to describe its force-displacement relationship.

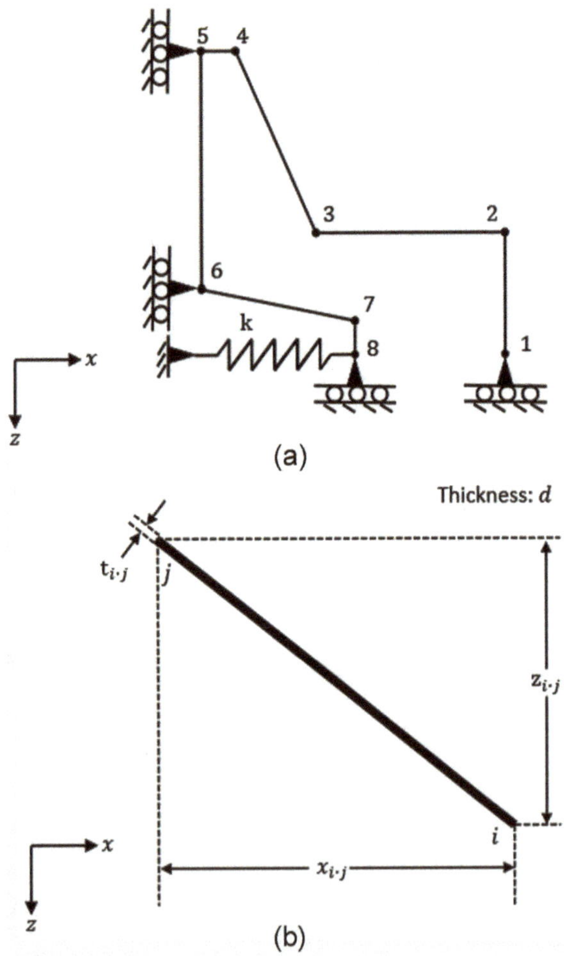

Figure 10.5 Illustration of (A) Schematic diagram of the two-stage force amplifier with eight nodes, and (B) Main parameters for flexure hinge *ij*.

Assume that the connecting linkages are flexure hinges with a small deformation and every connecting linkage is associated with two nodes. In this way, eight nodes are assigned for the TSFA. The input and output nodes are denoted as numbers 1 and 8, respectively. Nodes 1 to 8 are also used in Fig. 10.5A for reference. The piezoelectric stack is regarded as a spring.

The governing equation of the TSFA can be expressed in Eq. (10.2), which is derived by Hooke's law:

$$\boldsymbol{F} = \boldsymbol{K}\boldsymbol{U} \tag{10.2}$$

where K is the global stiffness matrix. F and U are the external force and displacement for all nodes, which can be expressed as:

$$\boldsymbol{F} = \left[(\boldsymbol{F}_1)^T \quad \cdots \quad (\boldsymbol{F}_i)^T \quad \cdots \quad (\boldsymbol{F}_n)^T \right] \tag{10.3}$$

$$\boldsymbol{U} = \left[(\boldsymbol{U}_1)^T \quad \cdots \quad (\boldsymbol{U}_i)^T \quad \cdots \quad (\boldsymbol{U}_n)^T \right] \tag{10.4}$$

where n is the total number of nodes. In Fig. 10.6, the force and displacement vectors of node i are expressed by $\boldsymbol{F}_i = \left[F_{ix} \quad F_{iz} \quad M_{iy} \right]$ and $\boldsymbol{U}_i = \left[x_i \quad z_i \quad \theta_{iy} \right]$, respectively. In addition, F_{ix}, F_{iz}, and M_{iy} are the force components of node i, and x_i, z_i, and θ_{iy} are the displacement components of node i.

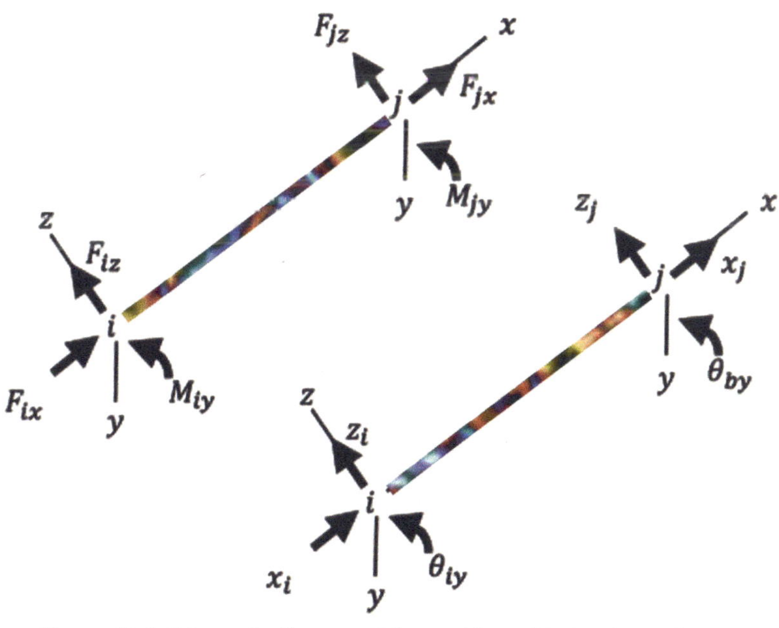

Figure 10.6 Schematic diagram of flexure hinge *ij* in local coordinates.

The global element stiffness matrix of flexure hinge ij $(\boldsymbol{K}_{ij})$ can be obtained as:

$$\boldsymbol{K}_{ij} = \boldsymbol{T}_{ij}(\boldsymbol{\alpha})\boldsymbol{K}_{ij}^{L}\left[\boldsymbol{T}_{ij}(\boldsymbol{\alpha})\right]^{T} \tag{10.5}$$

where $\boldsymbol{T}_{ij}(\boldsymbol{\alpha})$ and $\boldsymbol{K}_{ij}^{L}$ are the rotational transformation matrix and the local element stiffness matrix of flexure hinge ij, which can be expressed as:

$$\boldsymbol{T}_{ij}(\boldsymbol{\alpha}) = \begin{bmatrix} \cos\alpha & \sin\alpha & 0 & 0 & 0 & 0 \\ -\sin\alpha & \cos\alpha & 0 & 0 & 0 & 0 \\ 0 & 0 & 1 & 0 & 0 & 0 \\ 0 & 0 & 0 & \cos\alpha & \sin\alpha & 0 \\ 0 & 0 & 0 & -\sin\alpha & \cos\alpha & 0 \\ 0 & 0 & 0 & 0 & 0 & 1 \end{bmatrix} \tag{10.6}$$

$$\boldsymbol{K}_{ij}^{L} = \begin{bmatrix} \frac{EA}{l} & 0 & 0 & \frac{EA}{l} & 0 & 0 \\ 0 & \frac{12EI}{l^3} & \frac{6EI}{l^2} & 0 & -\frac{12EI}{l^3} & \frac{6EI}{l^2} \\ 0 & \frac{6EI}{l^2} & \frac{4EI}{l} & 0 & -\frac{6EI}{l^2} & \frac{2EI}{l} \\ -\frac{EA}{l} & 0 & 0 & \frac{EA}{l} & 0 & 0 \\ 0 & -\frac{12EI}{l^3} & -\frac{6EI}{l^2} & 0 & \frac{12EI}{l^3} & \frac{6EI}{l^2} \\ 0 & \frac{6EI}{l^2} & \frac{2EI}{l} & 0 & \cdot\frac{6EI}{l^2} & \frac{4EI}{l} \end{bmatrix} \tag{10.7}$$

where α is the rotation angle (counterclockwise direction) from the local to global coordinate. E, A, l, and I denote the Young's modulus, cross-section area, length, and moment of inertia of flexure hinge ij, respectively.

The boundary conditions are assigned according to the essential and natural boundary limits, which can be expressed as:

$$F_{1z} = 0,\ F_{8z} = 0,\ F_{5x} = 0,\ F_{6x} = 0,\ F_{8x} = -kx_{8x}. \tag{10.8}$$

where k represents the equivalent stiffness, which is equal to one-third stiffness of the piezoelectric stack.

Thus, the force amplification ratio (η_{TSFA}) of the TSFA is calculated as:

$$\eta_{TSFA} = \frac{F_{8x}}{F_{1x}} \tag{10.9}$$

Therefore, the force amplification ratio (η) of the dual-axial piezoelectric energy harvester is given by:

$$\eta = \eta_{CTM} \cdot \eta_{TSFA} \tag{10.10}$$

10.2.3.1 Modeling of piezoelectric stack

Assume that a harmonic force is exerted on the piezoelectric stack and the electric circuit with resistive loads is employed. The induced complex voltage (V) is given by (Zhao and Erturk, 2014):

$$V = \frac{j\omega d_{33}^{eff} F_0}{j\omega C_p^{eq} + (1/R_l)} \tag{10.11}$$

where ω is the angular excitation frequency, F_0 is the amplitude of the harmonic force, R_l represents the load resistance, and d_{33}^{eff} and C_p^{eq} denote the effective piezoelectric constant and equivalent capacitance of the piezoelectric stack, respectively.

In Eq. (10.11), the maximum output power (P_{max}) and corresponding best resistance ($R_{l,best}$) can be determined as:

$$P_{max} = \frac{\omega \left(d_{33}^{eff} F_0\right)^2}{2C_p^{eq}} \tag{10.12}$$

$$R_{l,best} = \frac{1}{\omega C_p^{eq}} \tag{10.13}$$

10.3 Parameter optimization

In this section, design parameters for the energy harvester device are optimized.

10.3.1 Selection of variable parameters

First, key parameters governing the performance of the underfloor energy harvester were identified. Eq. (10.1) shows that the force amplification ratio of the CTM was influenced by angle β and the friction coefficients. For easy operation and manufacturing, β was selected to be 45 degrees. To reduce friction, a lubricant was applied. Therefore, the friction coefficient was estimated at less than 0.2. That is, the friction coefficient was between 0 and 0.2, which provided a force amplification ratio of 0.23 to 0.5 for the CTM.

The parameters of flexure hinges, which most influence the performance of the TSFA, include the length, cross-section area, and inclination angle. The parameters of thick beams are fixed in the optimization, because they have little influence. Moreover, the force amplification ratio increases with an increase in the length and decrease in the cross-section area and inclination angle. In practice, the force amplification ratio cannot attain an ultralarge value because the largest length is constrained by the compactness requirement and the smallest cross-section area is restricted by the manufacturing capability. In addition, thickness d was selected to be 10 mm in this work for easy machining. Therefore, the variable parameters of the flexure hinges were determined to be:

$$P = [x_{3\cdot 4},\ z_{3\cdot 4},\ t_{3\cdot 4},\ x_{6\cdot 7},\ z_{6\cdot 7},\ t_{6\cdot 7}] \tag{10.14}$$

where parameters $x_{i\cdot j}$, $z_{i\cdot j}$, and $ti\cdot j$ are shown in Fig. 10.5B.

10.3.2 Optimization setup and results

Before optimization, the relationships between the parameters of flexure mechanism were assigned to ensure that the structure would change as expected. In addition, the material of the flexure mechanism was chosen to be Al-7075 because of its high elasticity. Piezoelectric stacks (RP150/5*5/30, Harbin Soluble Core Tech Co., Ltd.) were used in this study. Table 10.1 shows the main specifications of the adopted piezoelectric stack. The optimization and FEA simulation were processed by ANSYS Workbench software, and the GA method was selected as the optimization algorithm.

In this study, the entire underfloor energy harvester was expected to bear a maximum weight of 100 kg. That is, the maximum input force of each dual-axial piezoelectric energy harvester was around 500 N, which corresponded to the force amplification ratio of 0.5 for the CTM. Hence,

Table 10.1 Main parameters of piezoelectric stack.

Property	Symbol	Value
Dimension	$a \times a \times l$	5 × 5 × 28 mm
Stiffness	k_{PZT}	48.26 N/μm
Capacitance	C_p^{eq}	2.40 μF
Coefficient	d_{33}^{eff}	1.17×10^{-7} C/N

the magnitude of input force for the TSFA was 500 N. Both ends of piezoelectric stack were treated as fixed supports, and the magnitude of the reaction force for the fixed supports was measured, which was the force exerted at the piezoelectric stack. According to this setup, the analysis settings of the static structural analysis were implemented as shown in Fig. 10.7A.

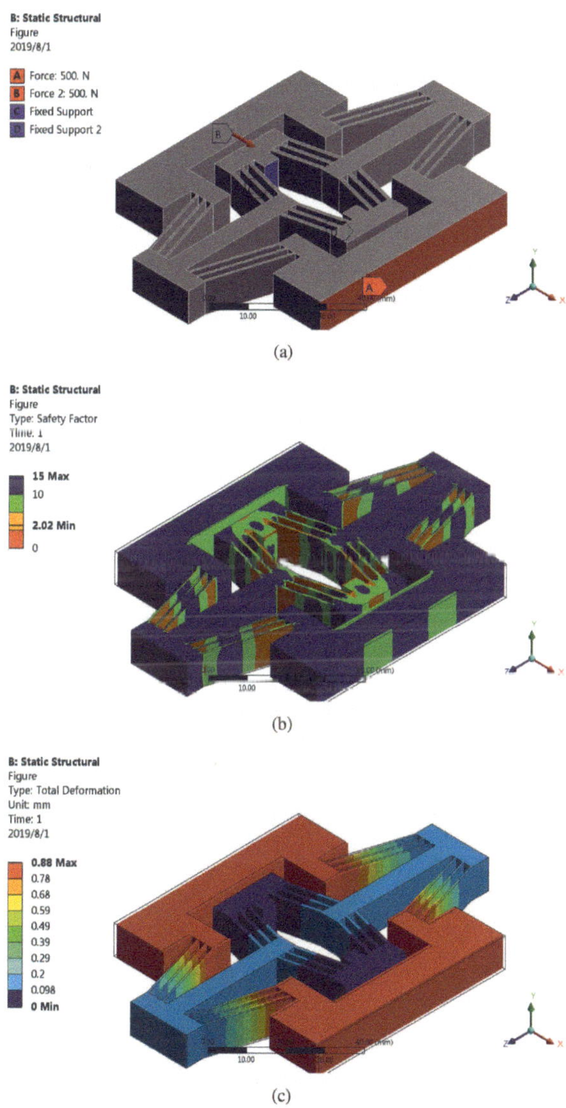

Figure 10.7 Simulation results of two-stage force amplifier. (A) Analysis setting; (B) Distribution of safety factor values; (C) Deformation result.

The primary objective of optimization is to improve force exerted at the piezoelectric stack to maximize energy output. To ensure that it could be used for a long time, a constraint was imposed such that the minimum safety factor was larger than 2.0. For easy machining and optimization, the minimum change of variable parameters was chosen to be 0.1 mm. The constraint set of the parameters is denoted as R^m.

The optimization function is:

- Objective: To maximize output force (P)
- Subject to:
  1. Minimum safety factor ≥ 2.0;
  2. $P \subseteq R^m$

Based on the optimization samples, the relationship of the output force versus the design parameters of outer flexure hinge 34 is shown in Fig. 10.8, which reveals that the output force decreases with a decrease in $z_{3 \cdot 4}$ or increase in $x_{3 \cdot 4}$ and $t_{3 \cdot 4}$. Based on these settings, the optimal parameters are obtained as listed in Table 10.2. When the TSFA is exerted by an input

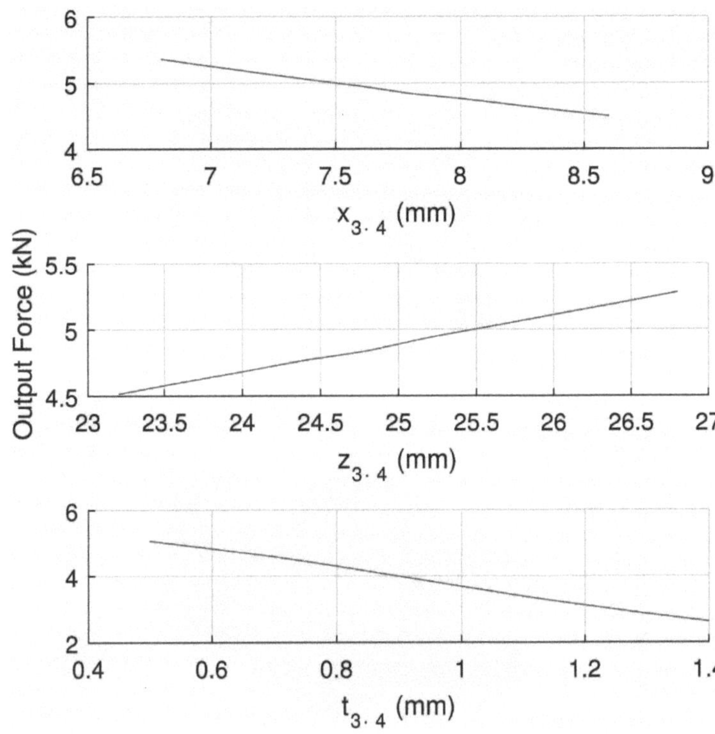

Figure 10.8 Relationship of output force versus design parameters for outer flexure hinge 34.

Table 10.2 Optimization results of harvester parameters.

Parameter	Value (mm)
$x_{3\cdot 4}$	7.8
$z_{3\cdot 4}$	24.8
$t_{3\cdot 4}$	0.6
$x_{6\cdot 7}$	14.9
$z_{6\cdot 7}$	3.7
$t_{6\cdot 7}$	1.0

force of 500 N, simulation results indicate that the embedded piezoelectric stack bears a force of 4838 N. Therefore, the force amplification ratios of the TSFA and dual-axial piezoelectric energy harvester are derived as $\eta_{\mathrm{TSFA}} = 9.68$ and $2.23 \leq \eta \leq 4.84$, respectively. Fig. 10.7B reveals that the minimum safety factor of the TSFA is 2.02, which ensures that it can be used for a long time for normal operations. Fig. 10.7C shows that the maximum deflection of input nodes for the TSFA is 0.88 mm, which is small enough and has no influence on human walking. Thus, the limited displacement of the limiters was set to 1.0 mm.

10.4 Prototype fabrication and experimental results

With the optimized structure parameters, a prototype underfloor energy harvester was fabricated for experimental verification. The details are presented in this section.

10.4.1 Prototype fabrication and experimental setup

The TSFAs were fabricated by wire-electrical discharge machining with a tolerance of ±0.05 mm. Other components were fabricated to satisfy functional demands. In particular, these components were fabricated by a stereolithography apparatus three-dimensional printer. The prototype and diagram of experimental setup are shown in Fig. 10.9. For experiments, the input force was supplied by a shaker (KSI-758ST150, KingSci Instruments, Inc.) and measured by a force sensor (LSB20, FUTEK Advanced Sensor Technology, Inc.). An electric circuit with resistive loads was employed. The piezoelectric stacks were connected to a variable resistance box (ZX21A, Shanghai Precision Instruments, Inc.). The input and output signals were processed by an NI PXIe-1082 real-time controller (National Instruments, Inc.). The hardware connection scheme is illustrated in Fig. 10.10.

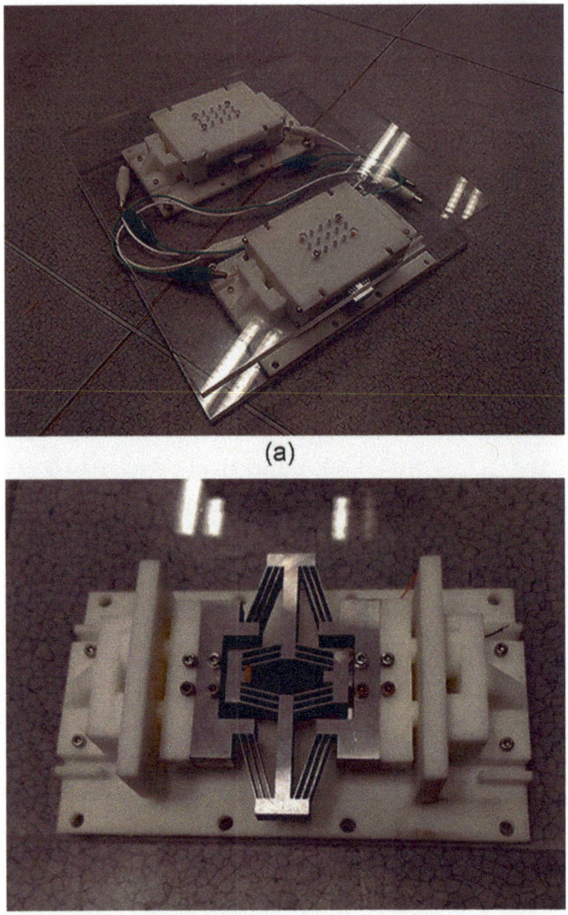

(a)

(b)

Figure 10.9 (A) The fabricated prototype underfloor energy harvester; (B) Inside view of fabricated two-directional piezoelectric energy harvester.

10.4.2 Identification of best load resistance

First, an experimental study was conducted to identify the best load resistance that could generate the maximum output power. The proposed underfloor energy harvester was considered to scavenge energy from humans walking at low frequency (0.8–1.2 Hz) (Ylli et al., 2015). Owing to the hardware constraint, the excitation frequency was selected to be 5 Hz, i.e. (the minimum excitation frequency of the shaker that was used). For illustration, the amplitude of the input force was selected to be 100 N. Fig. 10.11 shows the average output power with various load resistances. The power was obtained before rectifying the signal.

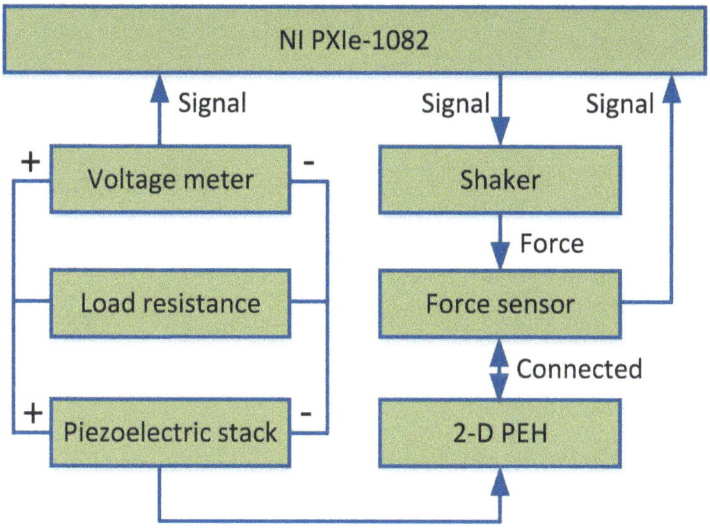

Figure 10.10 Connection scheme of experimental setup.

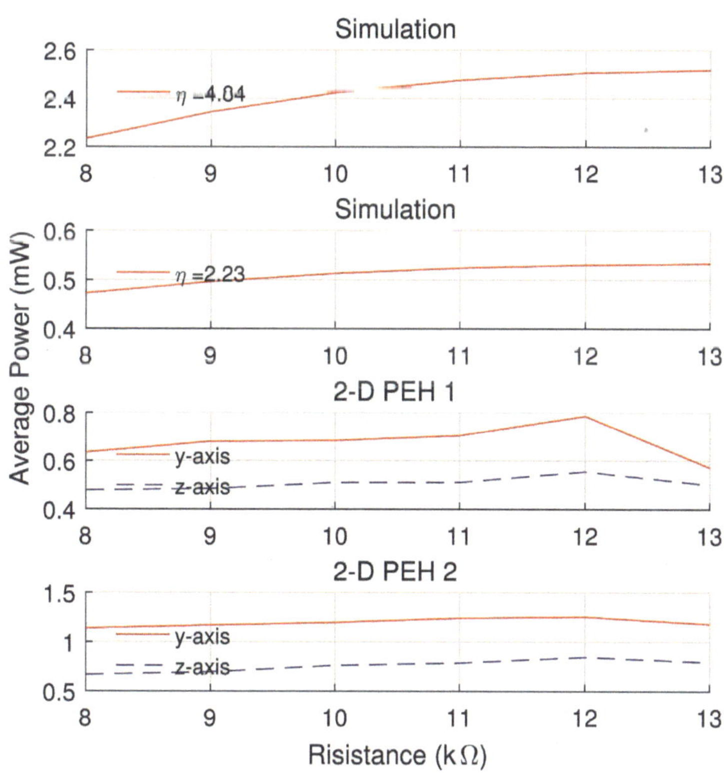

Figure 10.11 Experimental results of average output power versus load resistance for two dual-axial piezoelectric energy harvesters.

Table 10.3 Experimental results of maximum average output power and peak–peak output voltages.

Performance		Average output power (mW) at 12 kΩ	RMS output voltage (V)	Corresponding force amplification ratio
Two-directional piezoelectric energy harvester 1	y-axis	0.79	3.07	2.71
	z-axis	0.56	2.66	2.35
Two-directional piezoelectric energy harvester 2	y-axis	1.25	3.96	3.50
	z-axis	0.85	3.24	2.86
Simulation	$\eta_{CTM} = 0.5$	2.51	5.48	4.84
	$\eta_{CTM} = 0.23$	0.53	2.52	2.23

RMS, Root mean square.

Fig. 10.11 shows that the average output powers of the two tested dual-axial piezoelectric energy harvesters arrived at maximum values of around 12 kΩ under force applied in the y-axis or z-axis direction. Eq. (10.13) shows that under an excitation frequency of 5 Hz, the calculated best load resistance was 13 kΩ. Hence, the experimental result was close to the theoretical result with a small discrepancy of 7.7%.

Moreover, the average output powers at 12 kΩ are listed in Table 10.3. The simulation result of the average output power at 12 kΩ was between 0.53 and 2.51 mW. In contrast, the maximum average output power of the two piezoelectric energy harvesters was 0.79 and 1.25 mW, respectively, when the same input force was applied in the y-axis. In addition, the maximum average output power of the two piezoelectric energy harvesters was 0.56 and 0.85 mW, respectively, when the same input force was exerted in the z-axis. Therefore, all experimental results fell within the range predicted by the simulation study.

10.4.3 Performance test of fabricated dual-axial piezoelectric energy harvesters

Next, the performance of the two fabricated dual-axial piezoelectric energy harvesters was verified. With a 5-Hz sinusoidal input force of 100 N applied in the y- and z-axes, the voltmeter was connected in parallel with the corresponding best load resistance, which measured the

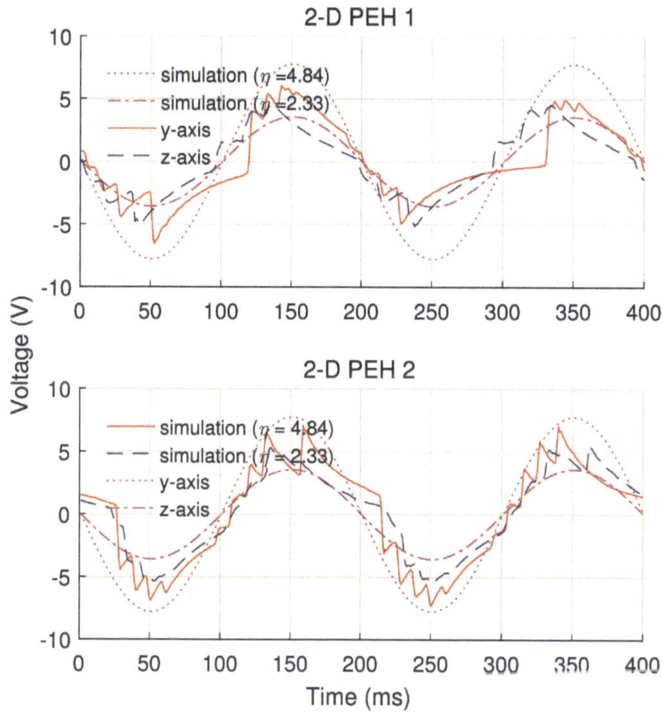

Figure 10.12 Experimental results of output voltage for the fabricated two dual-axial piezoelectric energy harvesters with the best load resistance.

average peak output power. The results of output voltage are shown in Fig. 10.12. Under the same input force, the experimental and simulation results of the output voltage for the fabricated dual-axial piezoelectric energy harvesters exhibited a similar tendency of variation. This demonstrates the feasibility of the proposed harvester design under 2-D input force excitation.

The RMS output voltages of the two fabricated piezoelectric energy harvesters are listed in Table 10.3. With the input force applied in the y-axis, the RMS output voltages of the two fabricated piezoelectric energy harvesters were 3.07 and 3.96 V, respectively. In addition, under the input force in the z-axis, the RMS output voltages of the two fabricated piezoelectric energy harvesters were 2.66 and 3.25 V, respectively. The simulation results for RMS voltage were 2.52−5.48 V, which well covered the experimental results. Moreover, the corresponding force amplification ratio of the experimental test was estimated as the ratio between the RMS output voltages of the experimental and simulation results multiplied by the simulation result for the force amplification ratio. The force amplification ratio of the dual-axial piezoelectric energy harvester was estimated to be 2.86, which was the mean value of the corresponding force amplification ratios of the experimental tests in Table 10.3.

Fig. 10.12 also reveals that the curves had periodic stagnation and intense jitter. There was a discrepancy between the experimental results of the y- and z-axes. Moreover, when the amplitude of the input force had a small value (less than 20 N) in the experimental study, the output voltage became much smaller than the simulation value. These problems may have arisen from fabrication errors, assembly errors, and the friction of some components (e.g., CTMs, slider mechanisms, and position limiters).

10.4.4 Practical application test of developed underfloor energy harvester

A practical application test of the developed underfloor energy harvester was conducted. According to Chen et al. (2017) and Masani et al. (2002), the formula of human walking force can be determined as:

$$F(t) = M \times [10\sin(3.918t + 0.1854) + 2.42\sin(16.5t - 0.8601)] \quad (10.15)$$

where M is the weight of the walker. In addition, because the friction coefficients of the walker's' shoes are hard to predict, the lateral friction force is not considered in the simulation. Thus, the calculated best resistance is approximately 25–106 kΩ. Therefore, the best load resistance was chosen to be 26 kΩ, which was found by a simulation. The output energy per walking footstep was derived from the average value of 10 tests, which were carried out by 10 people.

The output energy per walking footstep increases as the human weight increases. In Eq. (10.12), output energy is proportional to the square of human weight if the same speed and body geometry of walkers are assumed. Therefore, a quadratic trend curve with a zero-crossing point was fitted, as shown in Fig. 10.13, with a coefficient of 1.82 $\mu J/kg^2$ with $R^2 = 0.9444$. Moreover, the coefficient of the trend curve is 1.08 times that of the simulation curve, which is obtained without a lateral friction force. The discrepancy may have been caused by the effect of a lateral

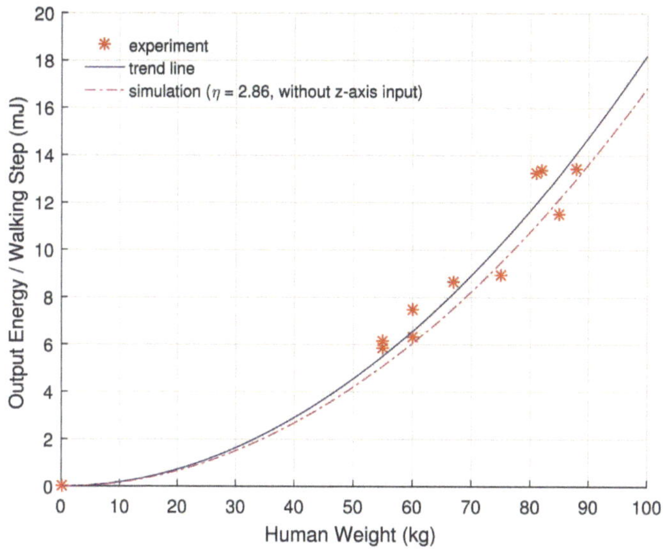

Figure 10.13 Experimental results for output energy per walking footstep at various human weights.

friction force. In addition, the average weight of people who were involved is around 62 kg. Thus, with an applied weight of 62 kg, the harvested average energy per walking footstep is 7.00 mJ for the proposed underfloor energy harvester, which is sufficient to power low-power consumption sensors.

Moreover, a study compared the proposed underfloor energy harvester and existing work in the literature (Gilbert and Balouchi, 2014; Puscasu et al., 2018; Zhu et al., 2013); and the results are shown in Table 10.4. The device in the existing work (Gilbert and Balouchi, 2014) offers much greater energy than others because of a larger input deflection, which may deteriorate the balance of walking people. Hence, it is unsuitable for use as an underfloor energy harvester. Among the other underfloor energy harvesters, the proposed underfloor energy harvester provides better performance than others in terms of harvested energy per walking footstep. Although the area harvested energy density and volume harvested energy density per walking footstep (Zhu et al., 2013) showed relatively large values, the data were provided by a small tested prototype rather than a whole prototype. Thus, the reported underfloor energy harvester is more suitable for scavenging energy from human walking.

Table 10.4 Performance comparison with existing work.

Performance	Proposed underfloor energy harvester	(Gilbert and Balouchi, 2014)	(Zhu et al., 2013)	(Puscasu et al., 2018)
Operating principle	Piezoelectric	Electromagnetic	Triboelectric	Piezoelectric
Applied weight (kg)	62	61	51–61	87
Maximum input deflection (mm)	1	10	1	1
Harvested energy per walking footstep (mJ)	7.00	450	0.74	2.44
Harvested energy per walking footstep (mJ)	7.00	450	0.74	2.44
Area harvested energy density per walking footstep (mJ/m^2)	78	–	1753	9.8
Volume harvested energy density per walking footstep (mJ/m^3)	1161	1.10×10^6	3.04×10^6	1394

10.5 Conclusion

This chapter conducted the design, optimization, fabrication, and testing of an underfloor energy harvester with two dual-axial piezoelectric energy harvesters. To satisfy the design objective, a flexure-based mechanism consisting of an TSFA, CTMs, slider mechanisms, and position limiters was devised. To maximize output performance, the structural parameters of the flexure mechanism were optimized by the GA method based on an FEA simulation under given constraints. A prototype harvester was fabricated and the performance of the fabricated dual-axial piezoelectric energy harvester was verified by experimental studies. Experimental results revealed that the developed dual-axial piezoelectric energy harvesters were feasible. They provide an average output power of 1.25−0.85 mW under a 5-Hz input force of 100 N applied in the y- and z-axes. With the developed underfloor energy harvester, the harvested energy per walking footstep was 7.00 mJ when the average weight of tested people was around 62 kg, which is sufficient to power low-power consumption sensors.

References

Chen, W., Wang, Y., & Deng, W. (2017). Deformable force amplification frame promoting piezoelectric stack energy harvesting: parametric model, experiments and energy analysis. *Journal of Intelligent Material Systems and Structures, 28*(7), 827−836. https://doi.org/10.1177/1045389X16657426

Gilbert, J. M., & Balouchi, F. (2014). Design and optimisation of a footfall energy harvesting system. *Journal of Intelligent Material Systems and Structures, 25*(14), 1746−1756. https://doi.org/10.1177/1045389X14523853

Masani, K., Kouzaki, M., & Fukunaga, T. (2002). Variability of ground reaction forces during treadmill walking. *Journal of Applied Physiology, 92*(5), 1885−1890. https://doi.org/10.1152/japplphysiol.00969.2000

Puscasu, O., Counsell, N., Herfatmanesh, M. R., Peace, R., Patsavellas, J., & Day, R. (2018). Powering lights with piezoelectric energy-harvesting floors. *Energy Technology, 6*(5), 906−916. https://doi.org/10.1002/ente.201700629

Saadatzi, M., Mir, F., Saadatzi, M. N., & Banerjee, S. (2018). Modeling and fabrication of a multi-axial piezoelectric energy harvester based on a metamaterial-inspired structure. *IEEE Sensors Journal, 18*(22), 9410−9419. https://doi.org/10.1109/JSEN.2018.2870993

Wang, L., Chen, S., Zhou, W., Xu, T. B., & Zuo, L. (2017). Piezoelectric vibration energy harvester with two-stage force amplification. *Journal of Intelligent Material Systems and Structures, 28*(9), 1175−1187. https://doi.org/10.1177/1045389X16667551

Wen, S., & Xu, Q. (2019). Design of a novel piezoelectric energy harvester based on integrated multistage force amplification frame. *IEEE, 24*(3), 1228−1237. https://doi.org/10.1109/TMECH.2019.2906999

Wen, S., Xu, Q., & Zi, B. (2018). Design of a new piezoelectric energy harvester based on compound two-stage force amplification frame. *IEEE Sensors Journal, 18*(10), 3989−4000. https://doi.org/10.1109/JSEN.2018.2820221

Wu, Z., & Xu, Q. (2019). Design and testing of a novel bidirectional energy harvester with single piezoelectric stack. *Mechanical Systems and Signal Processing, 122*, 139–151. https://doi.org/10.1016/j.ymssp.2018.12.026

Wu, Z., & Xu, Q. (2020). Design, fabrication and testing of a novel 3-D energy harvester with single piezoelectric stack. *Journal of Mechanical Design, 142*(6), 63303.

Yang, Z., & Zu, J. (2014). High-efficiency compressive-mode energy harvester enhanced by a multi-stage force amplification mechanism. *Energy Conversion and Management, 88*, 829–833. https://doi.org/10.1016/j.enconman.2014.09.026

Ylli, K., Hoffmann, D., Willmann, A., Becker, P., Folkmer, B., & Manoli, Y. (2015). Energy harvesting from human motion: Exploiting swing and shock excitations. *Smart Materials and Structures, 24*(2). https://doi.org/10.1088/0964-1726/24/2/025029

Zhao, S., & Erturk, A. (2014). Deterministic and band-limited stochastic energy harvesting from uniaxial excitation of a multilayer piezoelectric stack. *Sensors and Actuators A: Physical, 214*, 58–65. https://doi.org/10.1016/j.sna.2014.04.019

Zhu, G., Lin, Z. H., Jing, Q., Bai, P., Pan, C., Yang, Y., Zhou, Y., & Wang, Z. L. (2013). Toward large-scale energy harvesting by a nanoparticle-enhanced triboelectric nanogenerator. *Nano Letters, 13*(2), 847–853. https://doi.org/10.1021/nl4001053

CHAPTER 11

Design of a three-dimensional piezoelectric energy harvester

This chapter presents the design, fabrication, and testing of a piezoelectric energy harvester to harvest energy from three-degree-of-freedom (3-DOF) force excitation using a single piezoelectric stack. The piezoelectric energy harvester was constructed with several force transmission mechanisms and slider mechanisms. The best output performance of the energy harvester was achieved by selecting and optimizing the key parameters. A prototype was fabricated for experimental testing of the harvester performance. Results demonstrated the effectiveness of the developed 3-DOF piezoelectric energy harvester under the input force applied in three-dimensional space.

11.1 Introduction

Recently, a number of piezoelectric stack-based piezoelectric energy harvesters have been reported. However, majority of the researches are devoted to the force amplification frame, which only accepts unidirectional input force (Chen et al., 2015; Chen et al., 2017; Lee and Tovar, 2013; Wang et al., 2016; Wen et al., 2018). If an input force with multiple directions is applied, the energy produced by other directions of force cannot be scavenged. Normally, a multidirectional energy harvester is much practical under the situation with multidirectional excitation. Although a multidirectional energy harvester is easily realized by using multiple piezoelectric stacks, it is more economic if the energy harvester is developed with fewer piezoelectric stacks, such as using single piezoelectric stack to harvest energy from bidirectional excitation (Wu and Xu, 2019). From these perspectives, a single stack-based piezoelectric energy harvester dedicated to the situation with multidirectional excitation is desirable. The major challenge is how to use one piezoelectric stack to scavenge energy from 3-DOF input force.

Mechanical Design of Piezoelectric Energy Harvesters
ISBN 978-0-12-823364-1
https://doi.org/10.1016/B978-0-12-823364-1.00002-X
© 2022 Elsevier Inc.
All rights reserved.

This chapter presents a single stack-based piezoelectric energy harvester dedicated to the situation with 3-DOF excitation, such as underfloor harvesting on cross-road. In order to scavenge energy from 3-DOF input force, multiple force transmission mechanisms (FTMs) are introduced, which also realize the direction change of the input force. Multiple slider mechanisms (SMs) are adopted to eliminate redundant force components during the force direction transition. When there is a 3-DOF input force, the FTMs will change the input force into the direction which can be received by the force amplifier. In this work, a two-stage force amplifier (TSFA) is composed to amplify the force magnitude so as to improve the output power. The thin and thick beams have the characteristics of high force magnification ratio and high energy transmission ratio, respectively. The compound beam can effectively reduce the demanded thickness of each stage. Specifically, in order to get a high transmission efficiency and to adapt to the size requirement of the entire mechanism, a two-stage force amplifier with two-beam compound frame (with notch hinges) was designed. To improve the output performance, the dominant parameters of the piezoelectric energy harvester were optimized. A prototype has been fabricated, and the proposed concept design was verified by conducting several experimental studies.

11.2 Mechanism design

This section introduces the working principle and analytical modeling of the designed piezoelectric energy harvester. Fig. 11.1 shows the computer-aided design (CAD) model of the proposed energy harvester. Fig. 11.2 illustrates the working flowchart of the energy harvester.

11.2.1 Working principle

Referring to Figs. 11.1—11.2, the working principle of the designed 3-DOF energy harvesting mechanism is presented in the following. When the input end of the force transmission mechanisms receives an input force, a reaction force perpendicular to the sliding surface of the force transmission mechanisms will be generated. Then, the slider mechanism will eliminate redundant components of the reaction force. Meanwhile, the dominant component of the reaction force will be transmitted to the next mechanism.

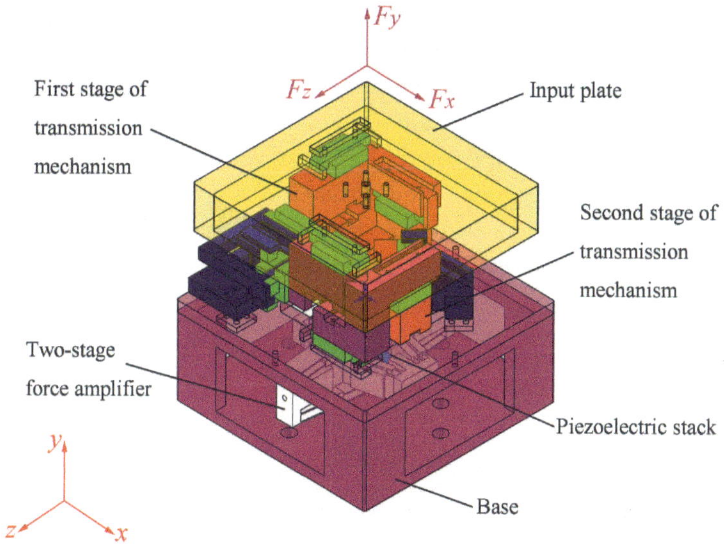

Figure 11.1 CAD model of the proposed piezoelectric energy harvester.

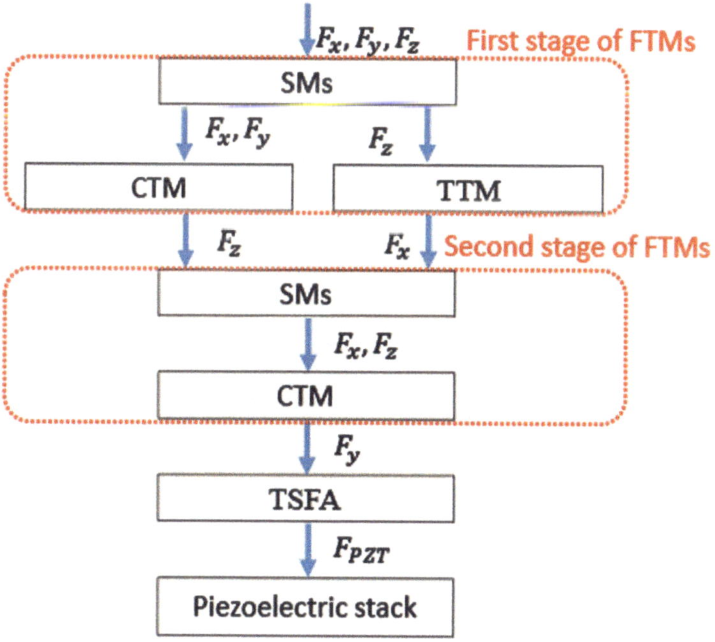

Figure 11.2 Working flowchart of the proposed piezoelectric energy harvester.

Specifically, when an input force with x-axis, y-axis, and z-axis components is applied on the proposed piezoelectric energy harvester, the x-axis and y-axis components will be transmitted as a force with only z-axis component after passing through the conical transmission mechanism (CTM) of the first stage of force transmission mechanisms. Then, the force with z-axis component will be transmitted into a force with only x-axis component after passing through the triangle transmission mechanism (TTM) of the first stage of force transmission mechanisms (see Fig. 11.3A).

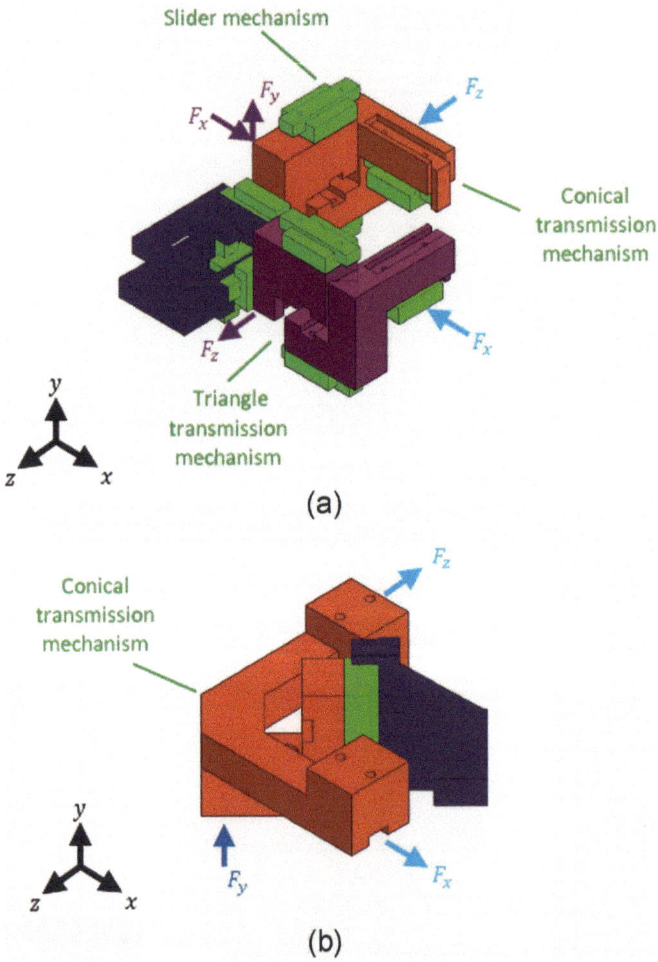

Figure 11.3 Illustration of (A) The first stage of force transmission mechanisms and (B) Second stage of force transmission mechanisms.

Afterward, the output force of the first stage of force transmission mechanisms with x-axis and z-axis components will be converted into a force with only y-axis component after passing through the conical transmission mechanism of the second stage of force transmission mechanisms (see Fig. 11.3B). Finally, the two-stage force amplifier will receive the force with y-axis component. The piezoelectric stack will experience the applied force after the force amplification via the two-stage force amplifier (see Fig. 11.4), which produces an electrical power based on direct piezoelectric effect.

11.2.2 Model of force transmission mechanism

The proposed piezoelectric energy harvester includes two kinds of force transmission mechanisms: conical and triangle. Although these two kinds of force transmission mechanisms have different shapes, their working principles are the same. In this study, the force relationship of the force transmission mechanism is derived by analytical modeling.

The static model of the force transmission mechanism is shown in Fig. 11.5. Because the structure is in static equilibrium, the force equation of the force transmission mechanism can be determined by:

$$F_{\text{in}} = R\sin\beta + \mu_{\text{FTM}} R\cos\beta + \mu_{\text{SM}} N_1 \tag{11.1}$$

$$R\cos\beta = \mu_{\text{FTM}} R\sin\beta + N_1 \tag{11.2}$$

$$N_2 = R\sin\beta + \mu_{\text{FTM}} R\cos\beta \tag{11.3}$$

$$R\cos\beta = F_{\text{out}} + \mu_{\text{FTM}} R\sin\beta + \mu_{\text{SM}} N_2 \tag{11.4}$$

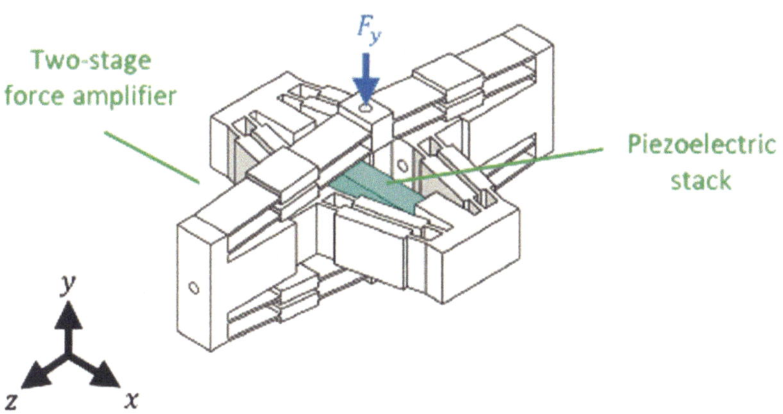

Figure 11.4 CAD model of the two-stage force amplifier with piezoelectric stack.

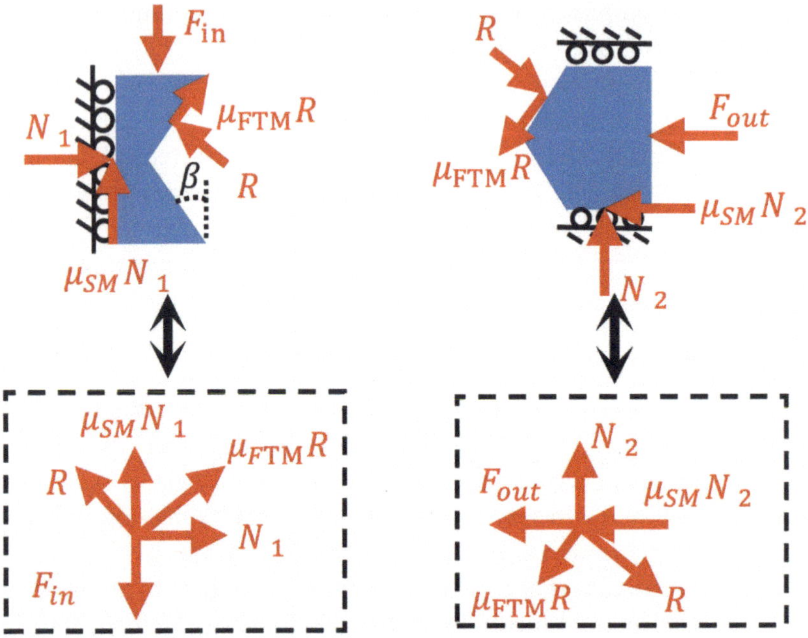

Figure 11.5 Force status of force transmission mechanism.

where R is the reaction force generated by the inclined plane; β is the angle of all inclined planes of the force transmission mechanisms, as shown in Fig. 11.5; F_{in} is the input force of the force transmission mechanism; F_{out} is the output force of the force transmission mechanism; N_1 and N_2 are the reaction forces generated by the slider mechanisms; and μ_{FTM} and μ_{SM} are the friction coefficients of the force transmission mechanism and slider mechanism, respectively.

In Eqs. (11.1)–(11.4), the force amplification ratio of the force transmission mechanism (η_{FTM}) can be derived as:

$$\eta_{FTM} = \frac{F_{out}}{F_{in}} = \frac{\cos\beta - \mu_{SM}\mu_{FTM}\cos\beta - (\mu_{FTM} + \mu_{SM})\sin\beta}{\sin\beta + \mu_{FTM}\cos\beta + \mu_{SM}(\cos\beta - \mu_{FTM}\sin\beta)} \quad (11.5)$$

Moreover, because it is hard to predict the friction coefficient, all sliding surfaces are assumed to be smooth and all friction coefficients are supposed to be zero. Therefore, the force amplification ratio of the force transmission mechanism with a zero friction coefficient ($\eta_{FTM,0}$) can be derived as:

$$\eta_{FTM,0} = \frac{\cos\beta}{\sin\beta} = \cot\beta \quad (11.6)$$

11.2.3 Model of two-stage force amplifier

In this study, the force–deflection relationship of the two-stage force amplifier is described by an analytical model. Because both stages of the two-stage force amplifier are symmetric, it is necessary only to model the flexure hinges of one stage of the structure. Assume that all connecting linkages are flexure hinges with a small deformation and each flexure hinge consists of two nodes. As shown in Fig. 11.6, there are four nodes in each stage of the two-stage force amplifier. The input nodes of both stages are denoted number 4. The output nodes of both stages are denoted number 1.

According to Hooke's law, the system equation of the stage can be expressed as:

$$\boldsymbol{F} = \boldsymbol{K}\boldsymbol{U} \tag{11.7}$$

where $\boldsymbol{K}$ is the global stiffness matrix, $\boldsymbol{F}$ is the external force, and $\boldsymbol{U}$ is the displacement. In addition, $\boldsymbol{F}$ and $\boldsymbol{U}$ for all nodes can be further written as:

$$\boldsymbol{F} = \left[(\boldsymbol{F}_1)^T \cdots \quad (\boldsymbol{F}_i)^T \quad \cdots \quad (\boldsymbol{F}_n)^T \right] \tag{11.8}$$

$$\boldsymbol{U} = \left[(\boldsymbol{U}_1)^T \quad \cdots \quad (\boldsymbol{U}_i)^T \quad \cdots \quad (\boldsymbol{U}_n)^T \right] \tag{11.9}$$

where the total number of the nodes is n and $\boldsymbol{F}_i$ and $\boldsymbol{U}_i$ are the force and displacement vectors of node i, which are expressed by:

$$\boldsymbol{F}_i = \left[F_{ix} \quad F_{iy} \quad M_{iz} \right], i = 1, 2, \cdots, n \tag{11.10}$$

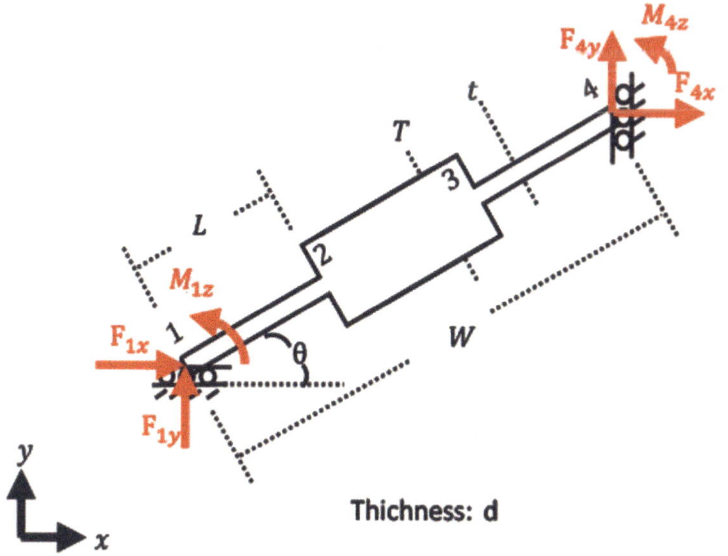

Figure 11.6 Schematic diagram of one branch of the two-stage force amplifier.

$$U_i = \begin{bmatrix} x_i & y_i & \theta_{iz} \end{bmatrix}, i = 1, 2, \cdots, n \tag{11.11}$$

in which the force components of node i are denoted by F_{ix}, F_{iy}, and M_{iz}; the displacement components of node i are denoted by x_i, y_i, and θ_{iz}, respectively.

In Fig. 11.7, the system equation of flexure hinge ij in the local coordinates can be further derived as:

$$F_i = K_{ij}^{11} U_i + K_{ij}^{12} U_j \tag{11.12}$$

$$F_j = K_{ij}^{21} U_i + K_{ij}^{22} U_j \tag{11.13}$$

where K_{ij}^{11} and K_{ij}^{22} have the same value; K_{ij}^{12} and $\left(K_{ij}^{21}\right)^T$ have the same value; and K_{ij}^{11} and K_{ij}^{12} can be expressed as:

$$K_{ij}^{11} = \begin{bmatrix} \frac{EA}{l} & 0 & 0 \\ 0 & \frac{12EI}{l^3} & \frac{6EI}{l^2} \\ 0 & \frac{6EI}{l^2} & \frac{4EI}{l} \end{bmatrix} \tag{11.14}$$

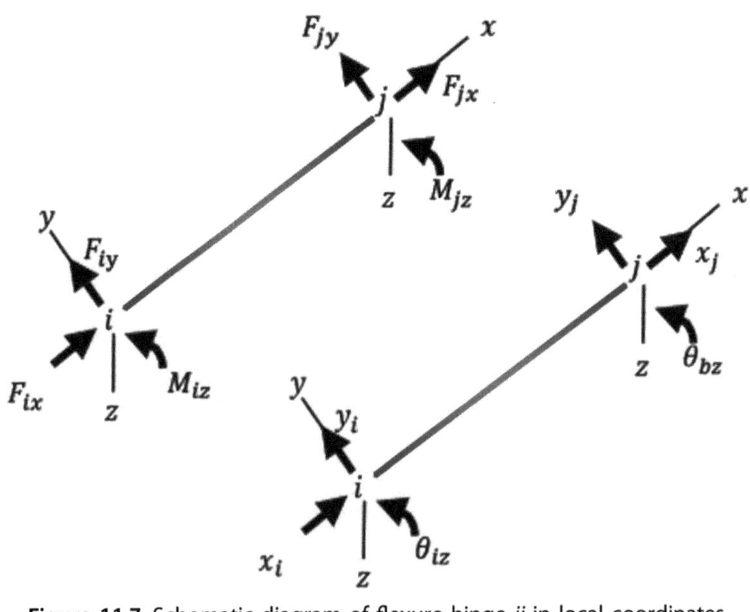

Figure 11.7 Schematic diagram of flexure hinge ij in local coordinates.

$$\boldsymbol{K}_{ij}^{12}=\begin{bmatrix}\dfrac{EA}{l} & 0 & 0\\ 0 & -\dfrac{12EI}{l^3} & \dfrac{6EI}{l^2}\\ 0 & -\dfrac{6EI}{l^2} & \dfrac{2EI}{l}\end{bmatrix} \tag{11.15}$$

where E denotes the Young's modulus; and l, A, and I represent the length, sectional area, and moment of inertia of flexure hinge ij, respectively.

Therefore, the local element stiffness matrix $\left(\boldsymbol{K}_{ij}^{L}\right)$ of flexure hinge ij can be determined as:

$$\boldsymbol{K}_{ij}^{L}=\begin{bmatrix}\boldsymbol{K}_{ij}^{11} & \boldsymbol{K}_{ij}^{12}\\ \boldsymbol{K}_{ij}^{21} & \boldsymbol{K}_{ij}^{22}\end{bmatrix} \tag{11.16}$$

The rotational transformation matrix, which is adopted to transform the local coordinate into the global coordinate, is determined by 6 × 6 matrix $\boldsymbol{T}_{ij}(\alpha)$ as:

$$\boldsymbol{T}_{ij}(\alpha)=\begin{bmatrix}\cos\alpha & \sin\alpha & 0 & 0 & 0 & 0\\ -\sin\alpha & \cos\alpha & 0 & 0 & 0 & 0\\ 0 & 0 & 1 & 0 & 0 & 0\\ 0 & 0 & 0 & \cos\alpha & \sin\alpha & 0\\ 0 & 0 & 0 & -\sin\alpha & \cos\alpha & 0\\ 0 & 0 & 0 & 0 & 0 & 1\end{bmatrix} \tag{11.17}$$

where α is the counterclockwise rotation angle from the local to global coordinate. Hence, the global element stiffness matrix $(\boldsymbol{K}_{ij})$ can be derived as:

$$\boldsymbol{K}_{ij}=\boldsymbol{T}_{ij}(\boldsymbol{\alpha})\boldsymbol{K}_{ij}^{L}\left[\boldsymbol{T}_{ij}(\boldsymbol{\alpha})\right]^{T} \tag{11.18}$$

Then, the global stiffness matrix of system ($\boldsymbol{K}$) can be expressed by:

$$\boldsymbol{K}=\begin{bmatrix}\ddots & & & & \iddots\\ & \boldsymbol{K}_{ii} & \cdots & \boldsymbol{K}_{ij} & \\ & \vdots & \ddots & \vdots & \\ & \boldsymbol{K}_{ji} & \cdots & \boldsymbol{K}_{jj} & \\ \iddots & & & & \ddots\end{bmatrix} \tag{11.19}$$

For the essential and natural constraints, the boundary conditions are expressed as:

$$F_{1x} = -F_{4x} \tag{11.20}$$

$$F_{1y} = -F_{4y} \tag{11.21}$$

$$M_{1z} = M_{4z} \tag{11.22}$$

$$F_{4y,\mathrm{outer}} = -|F_{\mathrm{TSFA}}|/4 \tag{11.23}$$

$$F_{1x,\mathrm{outer}} = F_{4y,\mathrm{inner}} \tag{11.24}$$

$$F_{1x,\mathrm{inner}} = -k_{PZT}|x_{1,\mathrm{inner}}|/2 \tag{11.25}$$

$$x_4 = y_1 = 0\ \mathrm{mm} \tag{11.26}$$

$$x_{4,\mathrm{outer}} = -y_{1,\mathrm{inner}} \tag{11.27}$$

where F_{TSFA} is the input force of two-stage force amplifier, k_{PZT} represents the stiffness of the piezo stack, subscript *inner* means that the parameter indicates the inner stage, and subscript *outer* indicates that the parameter belongs to the outer stage.

Therefore, the force amplification ratio of the two-stage force amplifier (η_{TSFA}) is given by:

$$\eta_{\mathrm{TSFA}} = \frac{|F_{1x,\mathrm{inner}}|}{|F_{4y,\mathrm{outer}}|} \tag{11.28}$$

The proposed piezoelectric energy harvester has two stages of force transmission mechanisms and one two-stage force amplifier. The amplification ratios of the two stages of force transmission mechanisms are identical. In Eqs. (11.6) and (11.28), the total force amplification ratio of the proposed piezoelectric energy harvester (η_{total}) can be calculated as:

$$\eta_{\mathrm{total}} = \eta^2_{\mathrm{FTM},0} \cdot \eta_{\mathrm{TSFA}} \tag{11.29}$$

11.2.4 Model of piezoelectric stack

Consider a harmonic force $F(t) = F_0 e^{-j\omega t}$ that acts directly on the piezo stack, where F_0 and ω are the amplitude and frequency of the harmonic force, respectively. According to Zhao and Erturk (2014), the complex voltage (V) is expressed by:

$$V = \frac{j\omega d_{33}^{eff} F_0}{j\omega C_p^{eq} + (1/R_l)} \tag{11.30}$$

where d_{33}^{eff} denotes the effective piezoelectric constant, C_p^{eq} is the equivalent capacitance, and R_l represents the resistance of load. Thus, the open-circuit voltage (V_{open}) and the output power (P) are given by:

$$V_{open} = \frac{d_{33}^{eff} F_0}{C_p^{eq}} \tag{11.31}$$

$$P = \frac{|V|^2}{R_l} = \frac{\left(\omega d_{33}^{eff} F_0\right)^2}{\left(\omega C_p^{eq}\right)^2 R_l + 1/R_l} \tag{11.32}$$

In Eq. (11.32), when $R_l = 1/\left(\omega C_p^{eq}\right)$, the maximum output power (P_{max}) will be obtained, which can be written as:

$$P_{max} = \frac{\omega\left(d_{33}^{eff} F_0\right)^2}{2C_p^{eq}} \tag{11.33}$$

11.3 Parameter optimization

This section provides the selection of design parameters, optimization setup, and optimization results. Based on finite element analysis (FEA) simulation, the multiobjective genetic algorithm (MOGA) method was applied to search for optimal parameters of the proposed piezoelectric energy harvester.

11.3.1 Selection of design parameters

To solve the optimization problem, the key design parameters that affect the output performance of the proposed piezoelectric energy harvester were selected based on the analytical modeling.

Eq. (11.6) reveals that angle β affected the performance of the force transmission mechanism. When angle β became larger, the force amplification ratio of the force transmission mechanism ($\eta_{FTM,0}$) was reduced. However, considering the reliability of manufacturing and subsequent tests, a force amplification ratio $\eta_{FTM,0} = 1$ was selected for the force transmission mechanism. Therefore, $\beta = 45$ degrees was derived. Hence, the input force of the two-stage force amplifier (F_{TSFA}) was expected to be equal to the input force of the entire energy harvester, which facilitated the later comparison study.

The parameters that affected the performance of the two-stage force amplifier included the length (l), sectional area (A), and counterclockwise rotation angle (α) from local to global coordinates of the flexure hinges. Because the thick beams can be regarded as rigid bodies, some parameters of the thick beams were not selected for optimization. Therefore, the main design parameters were determined as:

$$\left[l_{\text{inner}}, w_{\text{inner}}, h_{\text{inner}}, t_{\text{inner}}, d_{\text{inner}}, l_{\text{outer}}, w_{\text{outer}}, h_{\text{outer}}, t_{\text{outer}}, d_{\text{outer}} \right] \tag{11.34}$$

where the design parameters are shown in Fig. 11.8.

11.3.2 Optimization setup and results

Based on an FEA simulation with ANSYS Workbench software, the MOGA method was applied for optimization. The relationships between the parameters were determined before optimization to avoid unexpected changes in structure during optimization. In addition, Al-7075 was selected as the material for the flexure mechanism because it is highly elastic.

A static structural analysis of the two-stage force amplifier was conducted. To compare the results conveniently with those in the literature, 100 N was selected as the magnitude of input force for the proposed piezoelectric energy harvester (F_{input}). The acting force of the two-stage force amplifier (F_{TSFA}) was calculated as 100 N. Because the piezo stack features large stiffness, it is regarded as a rigid body in the simulation. Thus, the output force is a reaction force detected at the end of the piezo stack.

To improve the output performance, the key objective of optimization is to maximize the magnitude of the output force in the static structural

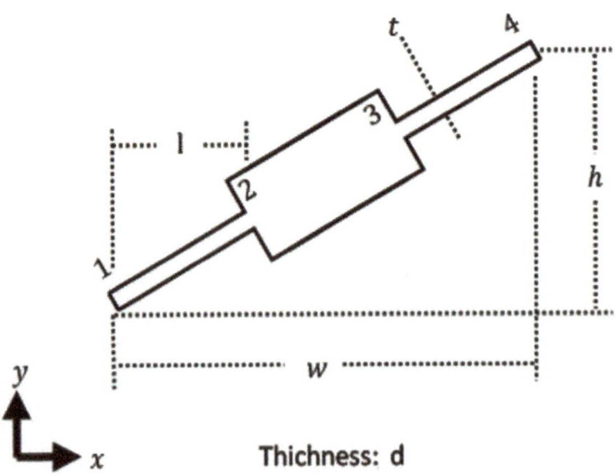

Figure 11.8 Variable parameters for flexure hinge *ij*.

analysis. Furthermore, minimizing the displacement of the input end of the two-stage force amplifier was selected as an objective of lower importance; it was expected to obtain good energy transmission efficiency. Generally, the minimum safety factor of a mechanism should be greater than 1.0. In this case study, a safety factor of 2.0 was chosen as the minimal requirement. For easy machining and optimization, a grid size of *d* and other parameters were selected as 1 and 0.1 mm, respectively. Also, the ranges of design parameters were defined in view of the total size of the mechanism design.

For the parametric study, the influence of each design parameter of the outer stage on the output force was obtained as shown in Fig. 11.9, which

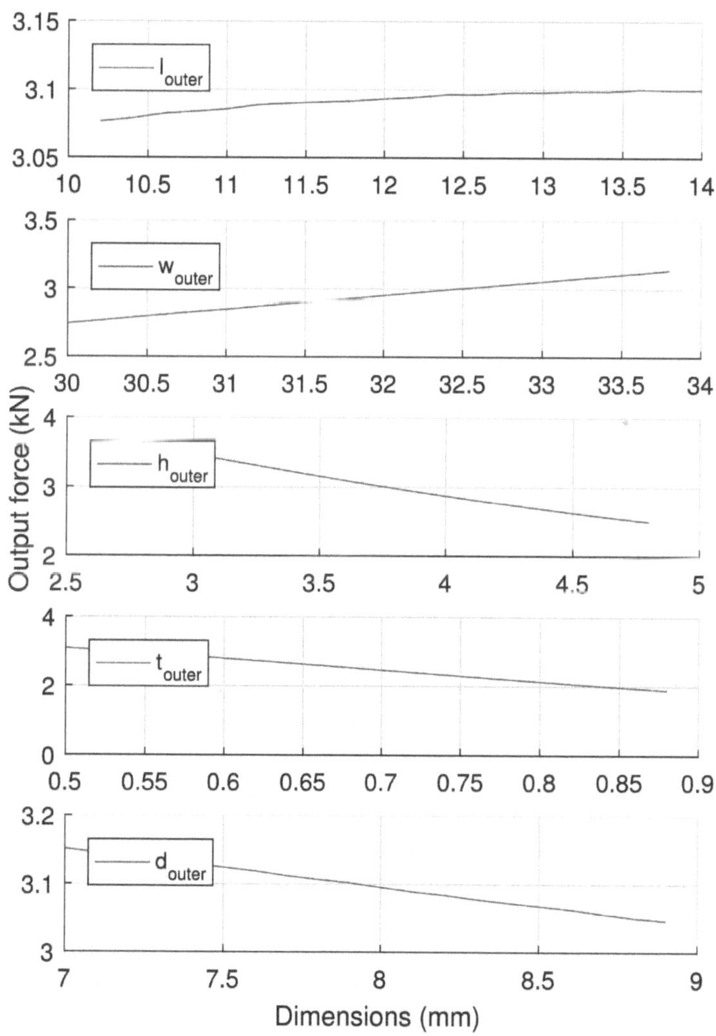

Figure 11.9 Influence of design parameters of outer stage on output force.

indicates that the output force increased with an increase in parameters *l* and *w* and a decrease in parameters *h*, *t*, and *d*. Thus, it was necessary to determine optimal parameters to achieve the maximum output force.

Table 11.1 shows the optimization results for the parameters based on the introduced constraints. The output force of the two-stage force amplifier was 3096 N under an input force of 100 N. Thus, the force amplification ratio of the two-stage force amplifier and the proposed piezoelectric energy harvester were both 30.96. The analysis settings for the static structural analysis are shown in Fig. 11.10A. In addition, Fig. 11.10B shows the deflection of the two-stage force amplifier, which indicates that the maximum input displacement of the two-stage force amplifier was 3.98 mm. Therefore, the heights of the force transmission mechanisms were selected to be 5.0 mm. Fig. 11.10C shows the simulation results for safety factor distribution for the two-stage force amplifier; the minimum safety factor for the two-stage force amplifier was 2.00. In addition, the energy transmission efficiency was equal to the ratio of the strain energies of the piezo stack and the total mechanism. Thus, the energy transmission efficiency was 33%.

11.4 Prototype fabrication and experimental results

A prototype harvester was fabricated with optimal parameters to verify the performance of the proposed piezoelectric energy harvester.

11.4.1 Prototype fabrication and experimental setup

Based on the CAD model, the two-stage force amplifier was manufactured from the selected material with a tolerance of ±0.05 mm. Some

Table 11.1 Optimization results of design parameters.

Parameter	Value (mm)
l_{inner}	3.8
w_{inner}	20.5
h_{inner}	4.7
t_{inner}	0.9
d_{inner}	10.0
l_{outer}	12.4
w_{outer}	33.4
h_{outer}	3.6
t_{outer}	0.5
d_{outer}	8.0

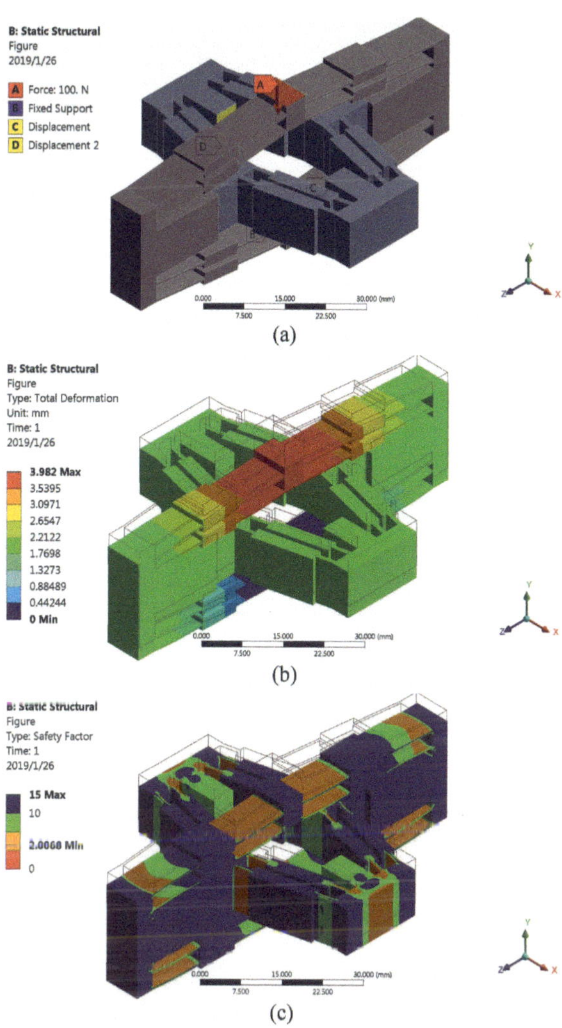

Figure 11.10 Finite element analysis simulation results for the two-stage force amplifier. (A) Analysis setting; (B) Deformation result; (C) Distribution of safety factor value.

components were manufactured by a stereolithography apparatus three-dimensional printer and others were designed for functional needs.

The fabricated prototype is shown in Fig. 11.11. The input force of the experiments was provided by a shaker (KSI-758ST150, KingSci Instruments, Inc.). A force sensor (LSB20, FUTEK Advanced Sensor Technology, Inc.) was adopted to measure the amplitude of input force.

Figure 11.11 Fabricated prototype of energy harvester.

A variable resistance box (ZX21A, Shanghai Precision Instruments, Inc.) was used to provide load resistance. A data acquisition board (NI PXIe-1082, National Instruments, Inc.) was used for data collection. A schematic diagram of the hardware connection is shown in Fig. 11.12. The main properties of the piezoelectric stack (RP150/5*5/30, Harbin Soluble Core Tech Co., Ltd.) are listed in Table 11.2.

The main purpose of the experimental studies was to verify the force magnification ratio. Because the bandwidth of the resonant circuit may

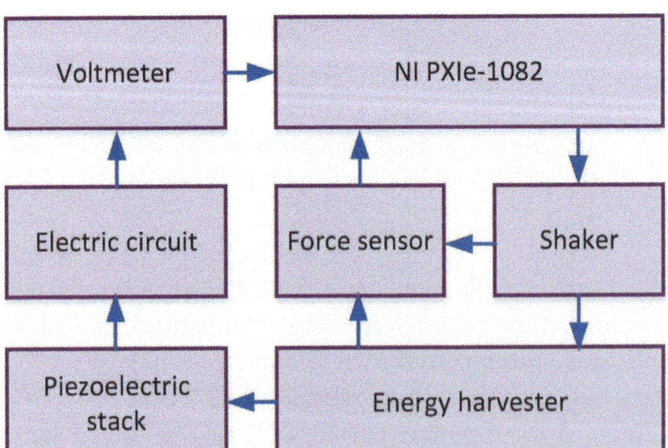

Figure 11.12 Scheme of hardware connection for energy harvester.

Table 11.2 Main properties of adopted piezoelectric stack.

Property	Symbol	Value
Dimension	$a \times a \times l$	$5 \times 5 \times 28$ mm
Stiffness	k_{PZT}	48.26 N/μm
Capacitance	C_p^{eq}	2.40 μF
Coefficient	d_{33}^{eff}	1.17×10^{-7} C/N

affect the output signal, an electric circuit with resistive loads was employed in the experimental studies.

11.4.2 Test results of load resistance effect

First, an experimental study was conducted to verify the influence of load resistance on average output power. Because the proposed piezoelectric energy harvester was mainly devoted to harvesting energy in low-frequency applications, such as foot strikes, resonance cases were not considered, because the first resonant frequency of the piezoelectric device was 175 Hz, according to FEA simulation results. For illustration, the average output power with various load resistances under a 5-Hz (i.e., minimum excitation frequency of the shaker) sinusoidal input force of 20 N is depicted in Fig. 11.13.

As shown in Table 11.2, the effective piezoelectric constant Cpeq of the piezoelectric stack was 2.4 μF. Thus, when the excitation frequency was 5 Hz, the optimal load resistance was calculated to be 13 kΩ. For the experimental results, when the proposed two-stage force amplifier or piezo stack received the input force directly, the maximum value of the average output power was achieved at 12 kΩ. In addition, when the input force in the x-, y-, and z-axes was applied to the harvester, the maximum values of the average output powers were also achieved at 12 kΩ (i.e., around the calculated optimal load resistance).

When the piezo stack received the input force directly, the peak value of average output power was 3.96 μW. By contrast, the peak value of average output power was 5137 μW with an input force acting on the proposed two-stage force amplifier, 468 μW under an input force in the x-axis, 249 μW under an input force in the y-axis, and 984 μW under an input force in the z-axis, which was 1297, 118, 63, and 248 times those of

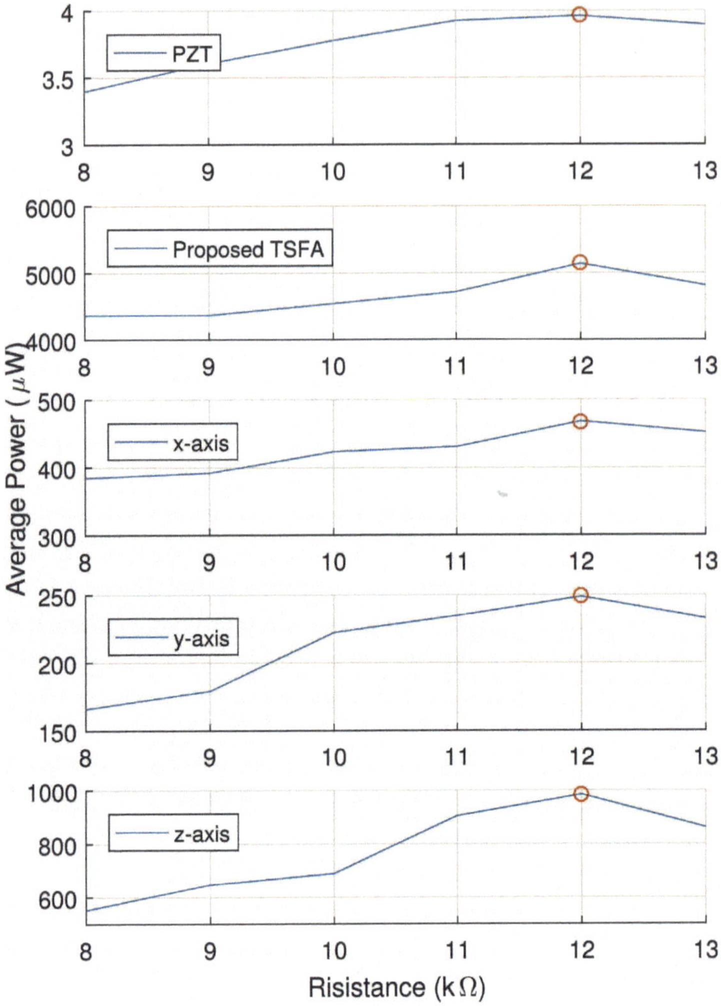

Figure 11.13 Experimental results of average output power with various load resistances under 5-Hz sinusoidal input force of 20 N.

the piezo stack, respectively. This shows that the output power was at a relatively low level, but it sufficed as an energy supplier for low-power micro-electromechanical systems devices, such as wireless sensors.

11.4.3 Test results for two-stage force amplifier

Next, the performance of the proposed two-stage force amplifier was verified by experimental study. The value of load resistance was chosen to be 12 kΩ, which is the corresponding best resistance of the input force with an excitation frequency of 5 Hz. Fig. 11.14 depicts the output voltage under a 5-Hz sinusoidal input force of 20 N.

The peak–valley output voltage reached 22.06 V under an input force directly acting on the two-stage force amplifier. In contrast, the peak–valley voltage of the simulation result was 19.87 V. The experimental results were compared with the simulation results, which revealed that the error of the force amplification ratio evaluation was 9.9%. This may induced a change in other performance, such as a reduction in the safety factor. Thus, the force amplification ratio of the two-stage force amplifier can be roughly derived as 34.37.

11.4.4 Test results for proposed energy harvester

The performance of the piezoelectric energy harvester was verified by experiments. The value of the load resistance was chosen to be 12 kΩ, which is the corresponding best resistance of the input force with an excitation frequency of 5 Hz. Fig. 11.15 depicts the output voltages under a 5-Hz sinusoidal input force of 20 N in the x-, y-, and z-axes.

The variation trends of the curves in Figs. 11.14 and 11.15 are similar, which reveals that the piezoelectric energy harvester can be adopted under the input force with 3-DOF direction. When the input force in the x-, y-,

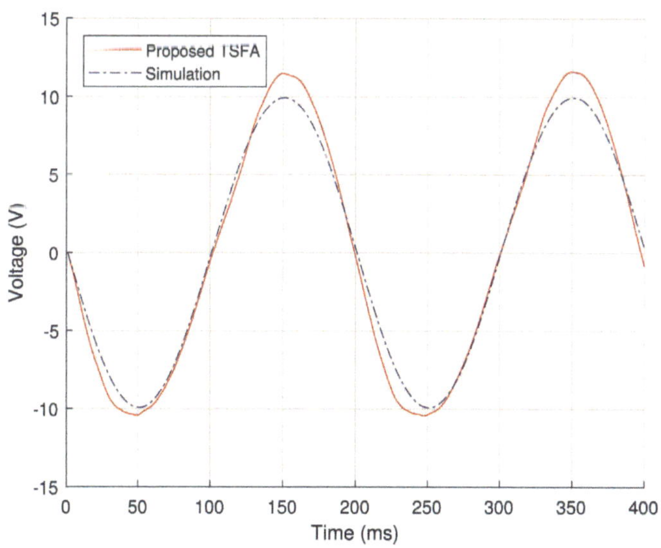

Figure 11.14 Experimental result of output voltage with 12 kΩ under 5-Hz sinusoidal input force of 20 N directly operated at proposed two-stage force amplifier.

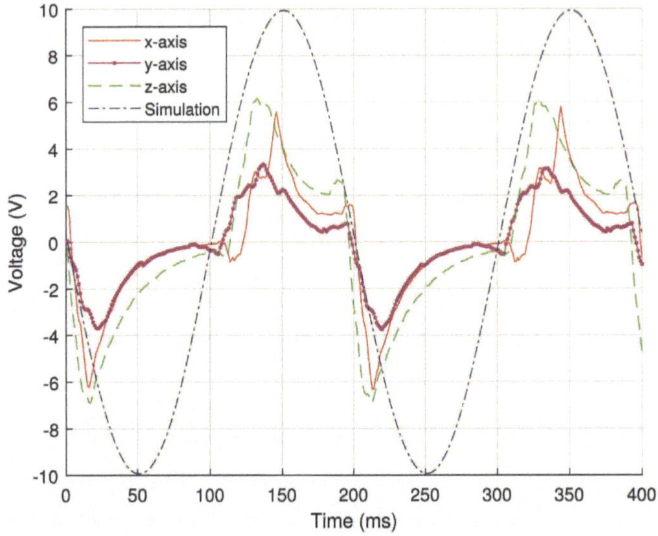

Figure 11.15 Experimental results of output voltage with load resistance of 12 kΩ under 5-Hz sinusoidal input force of 20 N in x-, y-, and z-axes.

and z-axes is exerted on the harvester, the peak—valley output voltages reach 12.12, 7.08, and 13.06 V, respectively. Compared with the simulation results, the experimental results under the input force in the x-, y-, and z-axes are 39.00%, 64.37%, and 34.27% lower than the simulation results, respectively. Hence, with the input force applied in x-, y-, and z-axes, the force amplification ratios are roughly 18.88, 11.03, and 20.35, respectively. Therefore, the force amplification ratios of the two stages of force transmission mechanisms in the x-, y-, and z-axes can be 0.55, 0.32, and 0.59, respectively, which are 0.45, 0.68, and 0.41 smaller than the expected value of 1.

The maximum deviation between the simulation and experimental results reached 64%. Moreover, the curve exhibited periodic stagnation and intense vibration. Additional experimental tests were manually conducted to verify the effectiveness of the piezoelectric energy harvester in different input directions, such as the direction with an angle of 45 degrees in the x-y plane. The experimental results revealed that the piezoelectric energy harvester was feasible in most directions. However, sometimes the piezoelectric energy harvester was not effective, such as when an input force in the y-axis acted on a location far from the sliding surfaces of the first stage of the force transmission mechanisms, where the input plate flipped over and could not return to its original position. These problems and the deviations of the force amplification ratio mainly came from fabrication errors, assembly errors, the bending deformation of the components, and friction in the mechanism. Future work will be carried out to overcome such issues.

11.4.5 Comparison with existing work

A performance comparison of the proposed design versus the existing design is provided in Table 11.3. To realize the presented concept design, the size of the proposed piezoelectric energy harvester became 5.94 times that of the two-stage force amplifier. For the proposed two-stage force amplifier and entire energy harvester, the simulation results for energy transmission efficiency, the safety factor, and the force amplification ratio were larger than those of existing work (Chen et al., 2015), which may have been because of its bigger size or different adopted material. With the input force applied in the x-, y-, and z-axes, the experimental results for the total force amplification ratio for the piezoelectric energy harvester were 10.1%, 47.5%, and 3.1% smaller than those of previous work (Chen et al., 2015), respectively. Even so, in the situation with 3-DOF excitation, the piezoelectric energy

Table 11.3 Performance comparison of piezoelectric energy harvesters in the literature.

Performance	Previous two-stage force amplifier (Wen et al., 2018)	Proposed two-stage force amplifier	Proposed piezoelectric energy harvester	
Material	60Si2CrA	Al-7075	Combined	
Size: L × W × H	72 × 45 × 25 mm	79.8 × 54 × 31.5 mm	94 × 94 × 91.3 mm	
Input force	100 N	100 N	100 N	
Energy transmission efficiency	18%	33%	33%	
Available directions	1-DOF	1-DOF	3-DOF	
Safety factor	1.6	2.00	2.00	
Amplification ratio (simulation)	24.4	30.96	30.96	
Amplification ratio (experiment)	21	34.37	x-axis	18.88
			y-axis	11.03
			z-axis	20.35

DOF, degree-of-freedom.

harvester generated more energy with higher energy transmission efficiency than in the existing work (Chen et al., 2015). The experimental results verified the feasibility of the proposed concept design of the 3-DOF piezoelectric energy harvester.

11.5 Conclusion

The design, optimization, and testing of a novel 3-DOF single stack-based piezoelectric energy harvester is reported in this chapter. To achieve the design objective using a single piezoelectric stack, a new mechanism was devised by integrating a two-stage force amplifier, force transmission mechanisms, and slider mechanisms. Design optimization for the structure parameters of the flexure mechanism was conducted to generate optimal output performance. To validate the actual performance of the proposed piezoelectric energy harvester, a prototype was fabricated and several experimental studies were conducted. The results of experiments show that the proposed concept design is practical. In future work, the performance of the energy harvester will be further improved by manufacturing with higher machining accuracy.

References

Chen, S., Wang, L., Zhou, W., Musgrave, P., Xu, T. B., & Zuo, L. (2015). Optimal design of force magnification frame of a piezoelectric stack energy harvester. In *Proceedings of SPIE - the international society for optical engineering* (Vol. 9435)SPIE. https://doi.org/10.1117/12.2084284

Chen, W., Wang, Y., & Deng, W. (2017). Deformable force amplification frame promoting piezoelectric stack energy harvesting: parametric model, experiments and energy analysis. *Journal of Intelligent Material Systems and Structures, 28*(7), 827–836. https://doi.org/10.1177/1045389X16657426

Lee, S., & Tovar, A. (2013). Topology optimization of piezoelectric energy harvesting skin using hybrid cellular automata. *Journal of Mechanical Design, 135*(3). https://doi.org/10.1115/1.4023322. Transactions of the ASME.

Wang, Y., Chen, W., & Guzman, P. (2016). Piezoelectric stack energy harvesting with a force amplification frame: modeling and experiment. *Journal of Intelligent Material Systems and Structures, 27*(17), 2324–2332. https://doi.org/10.1177/1045389X16629568

Wen, S., Xu, Q., & Zi, B. (2018). Design of a new piezoelectric energy harvester based on compound two-stage force amplification frame. *IEEE Sensors Journal, 18*(10), 3989–4000. https://doi.org/10.1109/JSEN.2018.2820221

Wu, Z., & Xu, Q. (2019). Design and testing of a novel bidirectional energy harvester with single piezoelectric stack. *Mechanical Systems and Signal Processing, 122*, 139–151. https://doi.org/10.1016/j.ymssp.2018.12.026

Zhao, S., & Erturk, A. (2014). Deterministic and band-limited stochastic energy harvesting from uniaxial excitation of a multilayer piezoelectric stack. *Sensors and Actuators, A: Physical, 214*, 58–65. https://doi.org/10.1016/j.sna.2014.04.019

CHAPTER 12

Design of a bistable piezoelectric energy harvester

This chapter presents the design, fabrication, and testing of a new bistable piezoelectric energy harvester dedicated to vibration energy harvesting in the direction of gravity. Bistability is realized by combining the elastic potential energy of a bridge-type compliant mechanism (BCM) and the gravitational potential energy of a proof mass, which simplifies the preloading mechanism design. As a result, its potential energy traps have less inconsistency under excitation in the gravity direction, which leads to easier skipping between the two energy wells. A parametric study was conducted to investigate the influence of the structural parameters on the harvester's performance. A static structural analysis was conducted to predict the actual performance of the proposed harvester, which was verified by conducting simulation studies. A prototype device was fabricated for experimental testing. In addition, a figure of merit was introduced for performance comparison with existing designs. Experimental results indicate that the proposed device is more suitable for scavenging energy in the gravity direction.

12.1 Introduction

The objective of vibration energy harvester design is to improve output power (Li et al., 2015; Moro and Benasciutti, 2010) and operational frequency bandwidth (L G H et al., 2014; Sharpes et al., 2015; Upadrashta and Yang, 2018). To broaden the frequency bandwidth of a harvester, a nonlinear piezoelectric vibration energy harvester can be employed to achieve larger output power. There are two major approaches to introducing nonlinearity into vibration energy harvesters: via structure-based mechanism designs and permanent magnets. A structure-based mechanism with nonlinear stiffness can provide a large frequency bandwidth; several nonlinear vibration energy harvesters have been designed (Cottone et al., 2012; Gu, 2011; Zhang et al., 2018). A survey of designs of bistable vibration energy harvesters was presented in Harne and Wang (2013). The

Mechanical Design of Piezoelectric Energy Harvesters
ISBN 978-0-12-823364-1
https://doi.org/10.1016/B978-0-12-823364-1.00010-9
© 2022 Elsevier Inc.
All rights reserved.

performance of a structure-based bistable vibration energy harvester can be tuned only by changing the buckling beam's length of the flexure mechanism if other parameters remain unchanged. The increase in magnitude of the potential energy traps is realized by increasing displacement between the two stable states. Thus, it is less flexible than a permanent magnet-based vibration energy harvester in terms of parameter adjustment. Intuitively, the performance of a permanent magnet-based vibration energy harvester is better than that of a structure-based bistable vibration energy harvester with a similar footprint size. However, the permanent magnets may affect the normal work of the electronic device. Hence, this work focuses on a structure-based bistable vibration energy harvester, which is preferred for practical application.

In reality, most vibration sources are along the vertical direction owing to the gravity effect, such as human walking. In the direction of gravity, because of the effect of gravitational potential energy, the magnitude of the two potential energy traps of the structure-based bistable vibration energy harvester becomes inconsistent, as illustrated in Fig. 12.1. The magnitude of the potential energy traps defines the ease of overcoming the potential energy trap. With a large inconsistency, the required input energy also increases to stabilize at a high energy orbit. Hence, a small inconsistency is best for practical use. Moreover, bistability could even disappear when a large proof mass is composed. Thus, existing structure-based bistable

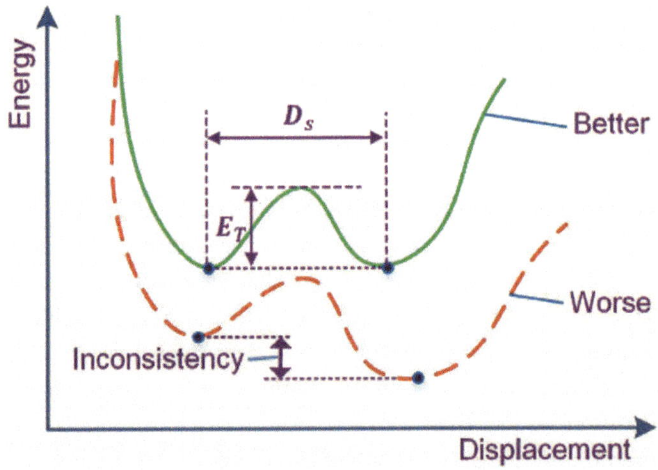

Figure 12.1 Energy–displacement relationship of a bistable vibration energy harvester. The two dots in each curve represent two stable positions of the bistable mechanism.

vibration energy harvesters are suitable only for harvesting energy in a direction without the effect of gravity. It is necessary to devise a structure-based bistable vibration energy harvester that is more suitable for energy harvesting dedicated to the direction of gravity.

This chapter reports the concept design of a structure-based bistable vibration energy harvester. It can scavenge vibration energy from the direction of gravity along with more balanced potential energy traps. Specifically, the proposed structure-based bistable vibration energy harvester is composed of two polyvinylidene fluoride (PVDF) films, a proof mass, and a BCM. Because of the buckling effect, the BCM functions as a nonlinear spring. Different from previous designs in which bistability was obtained by applying a preloading force (vertical to working axis) along the length direction of the buckling beams (Xu et al., 2013), the bistability of this design was generated by resorting to the buckling beams and gravity together. In addition, preloading of this design is imposed directly by gravity along the gravity direction, which simplifies the preloading mechanism design. Compared with previous designs of structure-based bistable vibration energy harvesters (C. Xu et al., 2013), the proposed structure-based bistable vibration energy harvester has less inconsistency when excitation is imposed in the direction of gravity, which is more suitable for harvesting energy in the gravity direction.

12.2 Analytical modeling

This section derives analytical models of the BCM and PVDF to evaluate the properties of the proposed structure-based bistable vibration energy harvester. Fig. 12.2 shows a computer-aided design (CAD) model of the designed structure-based bistable vibration energy harvester.

12.2.1 Model of bridge-type compliant mechanism

Because the BCM has a symmetric structure, the schematic diagram of the BCM is simplified, as shown in Fig. 12.3. The reaction force (F_{BCM}) and displacement (y) of the BCM along the y-axis are determined as:

$$F_{BCM} = 2F_S - mg \tag{12.1}$$

$$y = y_S \tag{12.2}$$

where F_S and y_S denote the reaction force and displacement of the simplified model along the y-axis, respectively. m represents the weight of the proof mass and g describes gravitational acceleration.

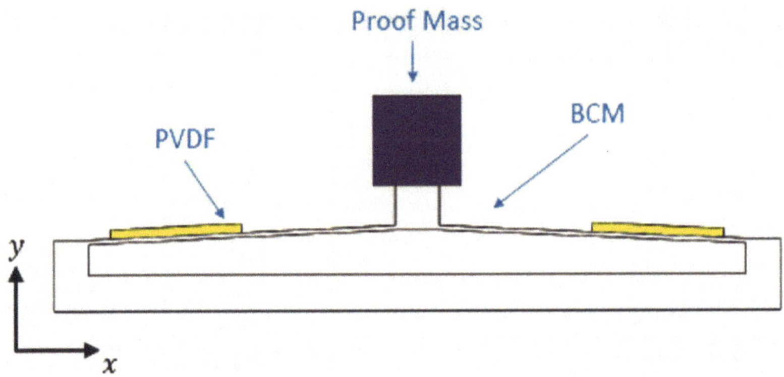

Figure 12.2 Computer-aided design model of proposed structure-based bistable vibration energy harvester.

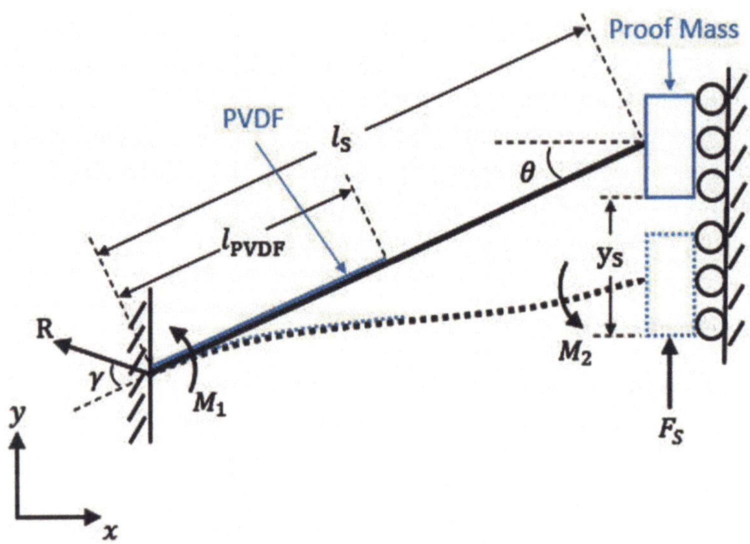

Figure 12.3 Simplified model of bridge-type compliant mechanism.

Because of the buckling effect, the stiffness of the BCM is nonlinear; it can be modeled using the elliptic integral approach. According to Howell et al. (2013) and Xu (2018), the governing equation of the flexure hinges is given as:

$$\sqrt{\eta} = F(\alpha, \beta_2) - F(\alpha, \beta_1) \tag{12.3}$$

$$1-\frac{y_S\sin\theta}{l_S}=-\frac{1}{\sqrt{\eta}}\{2\alpha\cos\gamma(\cos\beta_2-\cos\beta_1)\\+\sin\gamma[2E(\alpha,\beta_2)-2E(\alpha,\beta_1)-F(\alpha,\beta_2)+F(\alpha,\beta_1)]\} \tag{12.4}$$

$$1-\frac{y_S\sin\theta}{l_S}=-\frac{1}{\sqrt{\eta}}\{2\alpha\cos\gamma(\cos\beta_2-\cos\beta_1)\\+\sin\gamma[2E(\alpha,\beta_2)-2E(\alpha,\beta_1)-F(\alpha,\beta_2)+F(\alpha,\beta_1)]\} \tag{12.5}$$

where l_S and θ represent the length and inclined angle of the flexure hinges, respectively. Moreover, $\eta=\frac{Rl_S^2}{EI}$, in which R, E, and I denote the reaction force, equivalent Young's modulus, and moment of inertia of the composite hinges, respectively. In addition, γ is the force angle with respect to the flexure hinge at the original position. The first-kind and second-kind incomplete elliptic integrals are denoted by $F(\alpha,\beta)$ and $E(\alpha,\beta)$, which are given by:

$$F(\alpha,\beta)=\int_0^\beta\frac{d\delta}{\sqrt{1-\alpha^2\sin^2\delta}} \tag{12.6}$$

$$E(\alpha,\beta)=\int_0^\beta(\sqrt{1-\alpha^2\sin^2\delta})d\delta \tag{12.7}$$

where the modulus α is within the interval of [0, 1]. Variable β denotes the amplitude of the elliptic integral, which is the principal solution β_1 or the higher-order solution β_2 of:

$$\alpha\sin\beta_{1,2}=\cos\frac{\gamma}{2} \tag{12.8}$$

Thus, principal solution β_1 and higher-order solution β_2 can be determined as follows. The first mode of solution is given by:

$$\beta_1=\pi-\beta_2 \tag{12.9}$$

The second mode of solutions is given by:

$$\beta_2=\beta_1+2\pi \tag{12.10}$$

where the first mode and second mode of solutions represent deflections with a single inflection point and two inflection points, respectively.

Variables α, R, and γ can be obtained by solving the governing equations using a nonlinear solver. Then, the reaction force (F_S) of the simplified model along the y-axis can be calculated by:

$$F_S = R(\cos\gamma\sin\theta + \sin\gamma\cos\theta) \tag{12.11}$$

In Eqs. (12.1) and (12.11), the reaction force (F_{BCM}) of the BCM along the y-axis can be determined by:

$$F_{BCM} = 2R(\cos\gamma\sin\theta + \sin\gamma\cos\theta) - mg \tag{12.12}$$

Therefore, the total potential energy (U) of the proposed structure-based bistable vibration energy harvester can be derived as the sum of the elastic potential energy of the BCM and gravitational potential energy of the proof mass, which is given by:

$$U = \int_0^y (-F_{BCM})\mathrm{d}y \tag{12.13}$$

12.2.2 Model of proposed harvester

As shown in Fig. 12.2, the PVDFs are attached on the surface of the flexure hinges. The equivalent electromechanical model of the proposed structure-based bistable vibration energy harvester is given in Fig. 12.4. The governing equations of the motion are given by:

$$m\frac{\mathrm{d}^2y}{\mathrm{d}t^2} + c\frac{\mathrm{d}y}{\mathrm{d}t} - \frac{\mathrm{d}U}{\mathrm{d}y} + \phi V = -m\frac{\mathrm{d}^2u}{\mathrm{d}t^2} \tag{12.14}$$

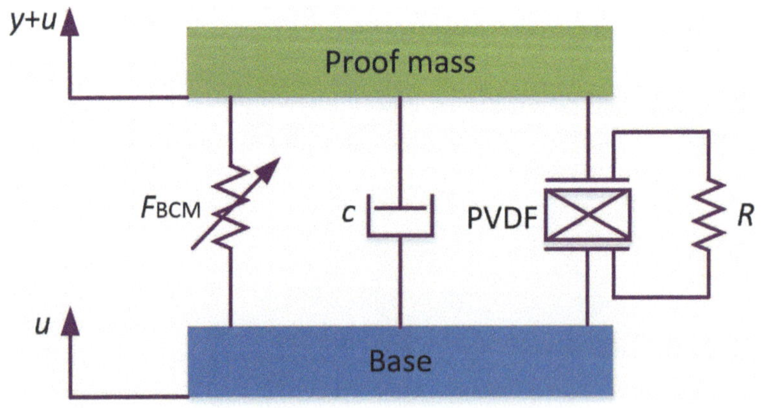

Figure 12.4 Equivalent electromechanical model of proposed structure-based bistable vibration energy harvester.

$$-\lambda\frac{dy}{dt}+\frac{dV}{dt}+\frac{V}{C_{PVDF}R}=0 \tag{12.15}$$

where c is the damping coefficient of the BCM; u is the base excitation displacement; ϕ and λ denote the equivalent piezoelectric coupling constant and the equivalent position-to-voltage coupling coefficient; V and R are the output voltage and load resistance, respectively; and C_{PVDF} is the capacitance of the PVDFs.

12.3 Parametric study and simulation verification

In this section, a parametric study is conducted based on the developed analytical models to predict the influence of the design parameters on the performance of the BCM.

12.3.1 Parametric study of bridge-type compliant mechanism

The material of the BCM is selected to be ABSplus-P430 (Stratasys Ltd., Eden Prairie, MN) with high elasticity. The three-dimensional printer (uPrint SE Plus, Stratasys Ltd.) is selected to be the fabrication machine. In addition, the PVDF (LDT1-028K, Measurement Specialties, Inc.) is adopted. Parameters of the fabrication material are described in Table 12.1. Parameters of the adopted PVDF and the equivalent model are described in Table 12.2. In addition, the equivalent piezoelectric coupling constant (ϕ) and equivalent position-to-voltage coupling coefficient (λ) are provided in Piezo Film Sensors Technical Manual (1999). Because nonlinearity is already considered in the restoring force model, the damping coefficient is assumed to be a small value (0.01) in the simulation. The value of the proof mass is selected to make the potential energy traps less inconsistent. The magnitude of the optimal proof mass (m_o), which produces the minimal inconsistence of the potential energy traps, is determined by trial-and-error.

Table 12.1 Parameters of material for prototype fabrication.

Parameter	Value
Density	1040 kg/m^3
Young's modulus	2.2 GPa
Poisson ratio	0.394
Yield strength	31 GPa

Table 12.2 Parameters of adopted polyvinylidene fluoride and equivalent model.

Parameter	Value
Length of polyvinylidene fluoride film	30 mm
Piezo strain constant (d_{31})	23 p.m./V
Piezo stress constant (g_{31})	216×10^{-3} Vm/N
Capacitance	1.38 nF
Young's modulus	3 GPa
Equivalent piezoelectric coupling constant (ϕ)	6.07×10^{-6} N/V
Equivalent position-to-voltage coupling coefficient (λ)	5.13×10^{3} V/m
Damping coefficient	0.01

In addition, the preliminary range of m_o can be roughly determined according to the negative stiffness region of F_S.

With the increase in the in-plane width and out-of-plane thickness of the flexure hinges, the stiffness and required proof mass also increase. To facilitate further operations and manufacturing, the structure-based bistable vibration energy harvester is considered to have a lighter proof mass. Therefore, the fabricated minimum in-plane width (1.0 mm) with the three-dimensional printer is assigned as the in-plane width of the flexure hinges. Moreover, the out-of-plane thickness of the flexure hinges is selected to be 16.5 mm, which is slightly larger than the dimension of PVDFs.

Furthermore, the influence of parameters l_S and θ on the performance of the BCM is shown in Fig. 12.5, which are generated by the derived analytical model. As shown in Fig. 12.5A and C, with an increase in l_S, m_o becomes smaller; the magnitude of the potential energy traps (E_T) shows a tendency to rise; the expected deflection (D_E) of PVDFs (which is equal to the product of displacement D_s between the two stable states and the length ratio between the piezo film of the PVDFs and flexure hinges) becomes larger.

As shown in Fig. 12.5B and D, with the increase in θ, the proof mass, magnitude of the potential energy traps, and expected deflection of PVDFs all become larger. In addition, the theoretical result is inaccurate when the resulted curve exhibits a sharp peak for $\theta = 3.5$ degrees, as shown in Fig. 12.5B. In this case study, to facilitate further testing, l_S and θ are selected to be 70 mm and 2.5 degrees, respectively. Hence, the theoretical value of the proof mass is derived as 54 g. The theoretical magnitude of the potential energy traps is 0.76 mJ, and the theoretical value of expected deflection of PVDFs is 1.87 mm.

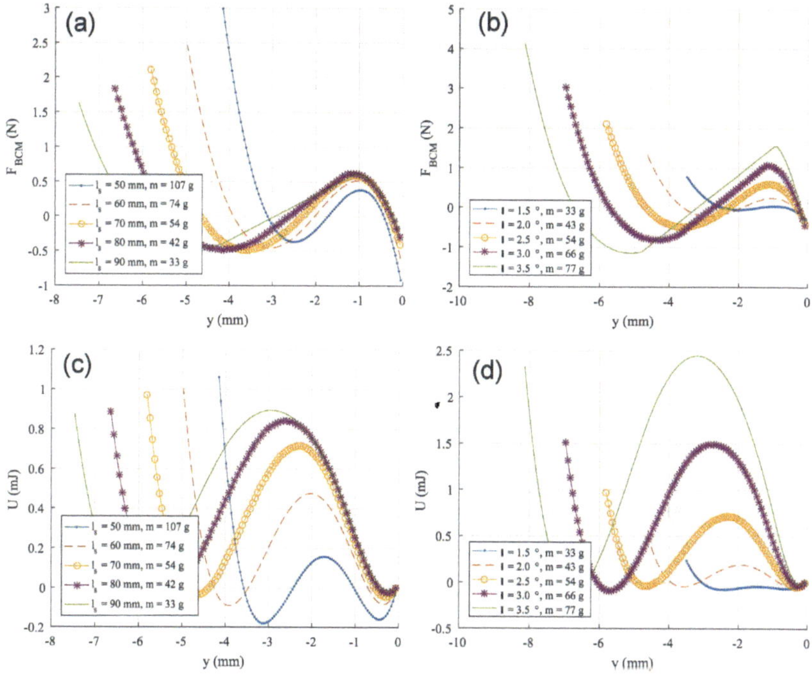

Figure 12.5 Theoretical performance of bridge-type compliant mechanism evaluated by analytical model. (A) and (C): Length of the flexure hinges is varied and $\theta = 2.5$ degrees; (B) and (D): Inclined angle of flexure hinges θ is varied and length is 70 mm.

12.3.2 Static structural analysis of bridge-type compliant mechanism

To evaluate reliability and verify the analytical model of the BCM, a static structural analysis was conducted based on finite element analysis (FEA) using ANSYS Workbench software.

The excitation end of the BCM was fixed and a variable displacement was loaded at the free end with the proof mass. To evaluate structural reliability, the minimum safety factor was measured. When the minimum safety factor was less than 1.0, the structure was unreliable because material failure occurred. To verify the analytical model of the BCM, the magnitude of the reaction force at the end with the proof mass was generated. In addition, because the BCM was nonlinear, the option of a large deformation was turned on during the simulation. Moreover, the mesh model was autogenerated with a relevance value of 100 and a relevance-center option of fine. Such a mesh led to a high accuracy of solution.

For illustration, the force-displacement and energy-displacement relationships obtained by the static structural analysis (by simulation) and the analytical model of the BCM are shown in Fig. 12.6A and B, respectively. For the analytical model, the mass was assigned to be 54 g, as discussed earlier, which produced minimal inconsistency of potential energy traps. As a result of the analytical modeling deviating from the FEA simulation, when the same proof mass (54 g) was assigned in the FEA simulation, results

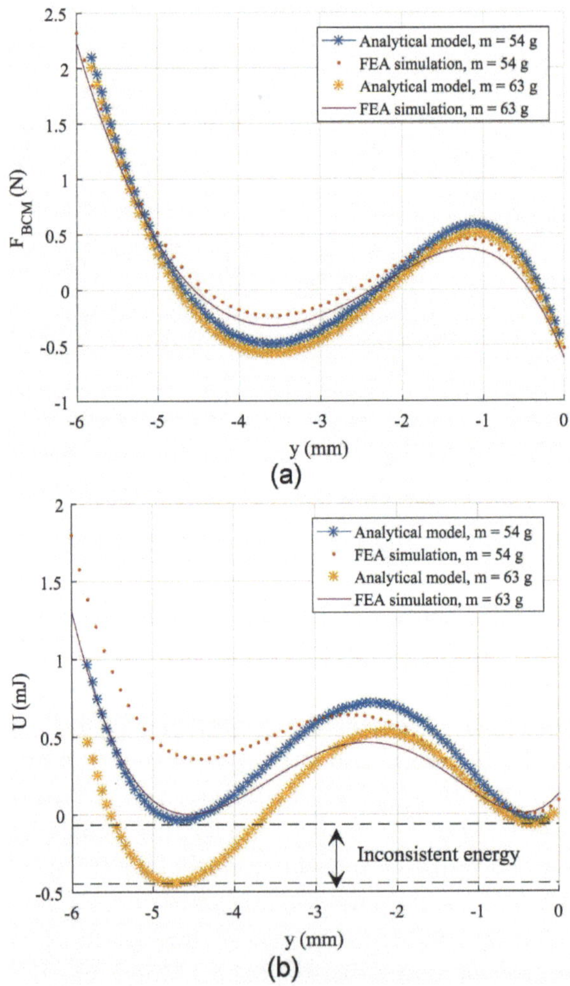

Figure 12.6 Analytical model and static structural analysis (simulation) results for (A) Force-displacement relationship and (B) Energy-displacement relationship of bridge-type compliant mechanism.

showed a relatively large inconsistency (i.e., 0.36 mJ) between the two potential energy traps. In addition, when the relatively inconsistent energy was twice as large as the magnitude of the potential energy traps without inconsistency, bistability disappeared. The inconsistent energy (U_i) in the direction of gravity could be predicted by the equation:

$$U_i = m_e g D_s \tag{12.16}$$

where m_e is the weight of the extra proof mass.

To produce the minimal inconsistency for the potential energy traps, the proof mass was tuned to be 63 g by trial-and-error. This obtained minimal inconsistency for potential energy traps, which was comparable to the analytical model results with a proof mass of 54 g. FEA simulation results also indicated that the magnitude of the potential energy traps was 0.46 mJ, and the expected deflection of PVDFs was 1.76 mm. Comparing the results of the analytical model and FEA simulation, there was some discrepancy between curves with different proof mass values. The slope changes of the force-displacement curves of analytical model result were larger than those of the FEA simulation result. This is why magnitudes of the potential energy traps of the analytical model were larger than those of the FEA simulation. The discrepancy was mainly caused by the simplification process of analytical modeling, in which the ends of flexure hinges with deflection are approximated as elliptic arcs.

In addition, the relationship between the minimum safety factor and displacement is shown in Fig. 12.7. To avoid material failure, the range of displacement of the BCM was set between −9 and 5.5 mm to make the minimum safety factor larger than 1.0. Therefore, a constraint was considered in the simulation study, in which displacement of the BCM did not exceed the set range. Moreover, such a constraint was imposed by setting position limiters in the prototype.

12.3.3 Simulation study of proposed harvester

To evaluate the output performance of the structure-based bistable vibration energy harvester, based on the results of the FEA simulation and analytical models, a further simulation study was conducted by MATLAB® software. For illustration, the structure-based bistable vibration energy harvester was considered to have a base excitation of 0.5 g with a resistor load of 0.1 MΩ.

Fig. 12.8 shows the phase portraits of the structure-based bistable vibration energy harvester under various frequencies with a proof mass of

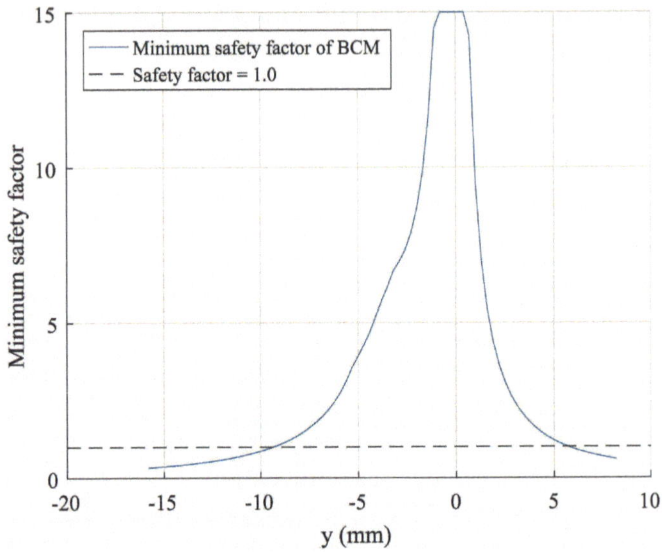

Figure 12.7 Relationship between minimum safety factor and displacement.

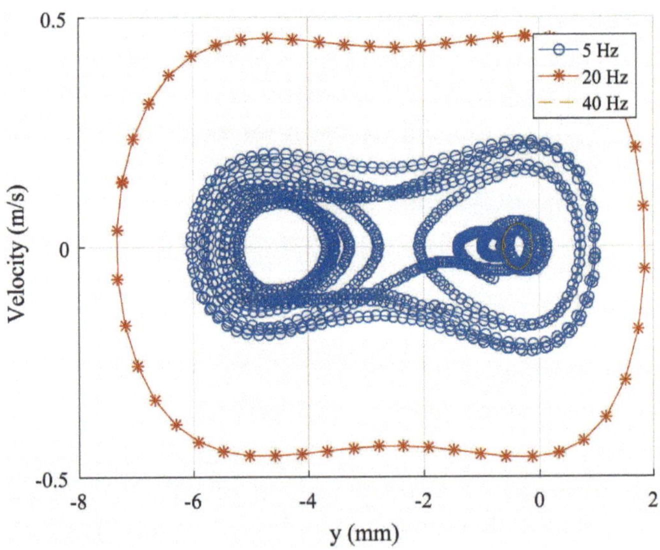

Figure 12.8 Phase portraits of proposed structure-based bistable vibration energy harvester under various frequencies with a proof mass of 63 g.

63 g. At a frequency of 40 Hz, the structure-based bistable vibration energy harvester was stabilized at a low energy orbit. The vibration energy harvester was confined in one of two potential traps and could not cross the energy barrier. At a frequency of 5 Hz, the vibration energy harvester oscillated between high and low energy orbits. The energy harvester repeated both motions of crossing the barrier and being confined in the potential trap. At a frequency of 20 Hz, the energy harvester was stabilized at a high energy orbit and there was no motion of being confined in the potential trap.

Fig. 12.9 shows the simulation results of average output powers with various frequencies and proof masses. In the simulation study with a proof mass of 63 g, the average output power reached a maximum value of 7.24 μW at a frequency of 20 Hz. The frequency range with over 30% peak value of average output power was around 13.1−20 Hz.

To evaluate the effect of inconsistency, simulation studies with proof masses of 43 and 83 g were also conducted. The inconsistent energies were both calculated as 0.81 mJ by Eq. (12.16). The frequency range with over a

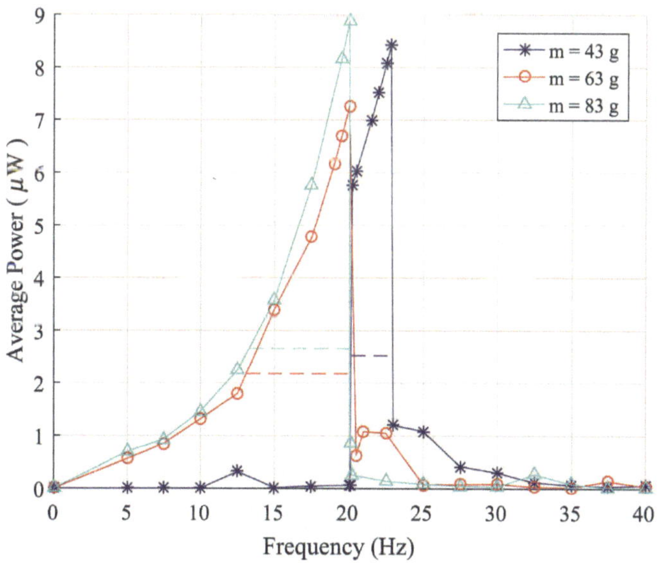

Figure 12.9 Simulation results of average output powers with various frequencies and proof masses. Dashed lines represent frequency ranges with 30% peak values.

30% peak value of output power under proof masses of 43 and 83 g was around 20.2–22.8 Hz and 13.3–20 Hz, respectively. The results indicate that the frequency range is reduced with an increase in inconsistent energy. Moreover, the discrepancy between the reduction of frequency ranges with proof masses of 43 and 83 g resulted from the difference in input energies. With the same base excitation, the input energy increased when the weight of the driving part increased.

12.4 Prototype fabrication and experimental results

In this section, a prototype of the proposed device was fabricated and the performance was verified by conducting several experimental studies.

12.4.1 Prototype fabrication and experimental setup

Based on the CAD model, the proposed structure-based bistable vibration energy harvester was manufactured with a tolerance of ± 0.2 mm. The PVDFs and BCM were connected by epoxy adhesive (3-ton F-05, Alteco Chemical Pte Ltd.). Owing to the direct piezoelectric effect, the PVDFs generate electric energy by deflecting flexure beams of the BCM. The experimental setup is shown in Fig. 12.10.

The input excitation of the experiments was provided by a shaker (KSI-758ST150, KingSci Instruments, Inc.). A laser displacement sensor (LK-H055, Keyence, Inc.) was adopted to measure the amplitude of excitation. A variable resistance box (ZX21A, Shanghai Precision

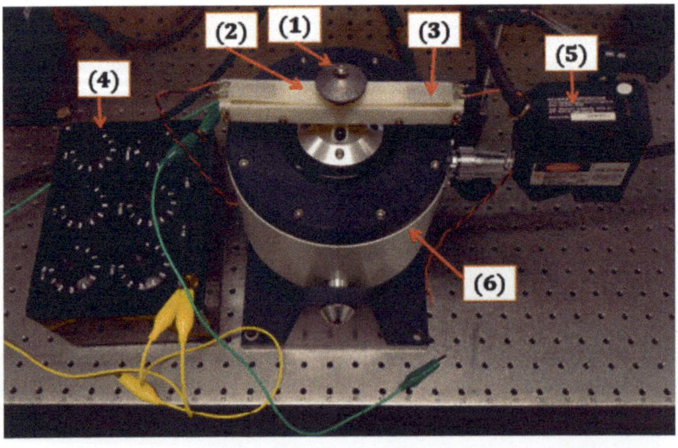

Figure 12.10 Experimental setup with (1) Proof mass, (2) Bridge-type compliant mechanism, (3) PVDF, (4) Resistance box, (5) Laser displacement sensor, and (6) Shaker.

Instruments, Inc.) was used to provide load resistance. A real-time controller (NI PXIe-1082, National Instruments, Inc.) was employed for data collection. Moreover, by measuring the weight of the proof mass and its fixtures, the equivalent proof mass in the experimental tests was derived as 65 g, which is close to the optimal value in the simulation. The base excitation and load resistance were 0.5 g and 0.1 MΩ, respectively, which were the same as those in the simulation setup.

12.4.2 Performance test results

An experimental study was conducted to test the performance of the proposed harvester at various frequencies. The relationship of the output power versus excitation frequency is shown in Fig. 12.11. The average output power reached its maximum value of 5.87 μW at 20 Hz. The frequency range with over 30% peak value of the average output power was around 16–20.4 Hz. Moreover, in the experiment, the states of the structure-based bistable vibration energy harvester with 5, 20, and 40 Hz were consistent with simulation results depicted in Fig. 12.8.

Compared with the simulation results, the experimental results of variation tendency for the average output power versus frequency were similar, whereas the amplitude of the experimental result was slightly smaller than

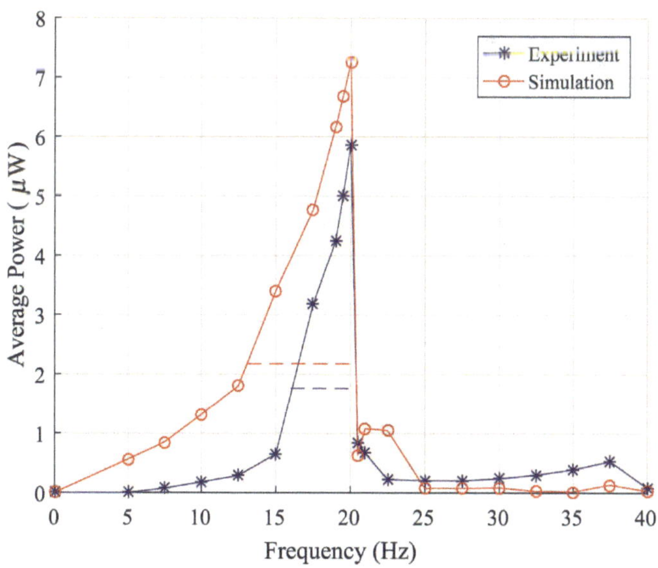

Figure 12.11 Generated average output power at various frequencies with a proof mass of 65 g.

that of the simulation result, as shown in Fig. 12.11. The discrepancy resulted from the manufacturing tolerance, different proof mass, different initial conditions, and simplification adopted in the modeling process. In addition, during the experiment, sometimes the proof mass swung left and right, which may have been caused by inconsistent stiffness of the left and right flexure hinges or assembly error. This was also a source of discrepancy between the experimental and simulation results.

12.4.3 Performance comparison study

For a comparison study, the performances of the bistable vibration energy harvester and typical structure-based bistable vibration energy harvesters in previous work (Andò et al., 2014; Xu et al., 2013) are listed in Table 12.3. The predicted inconsistent energy in the direction of gravity is derived by Eq. (12.16).

12.4.3.1 Proposal of figure of merit

To make a relatively fair comparison of the performances of available designs, a figure of merit is introduced as follows.

First, to rule out the effects of the performance difference of the flexure hinge, a flexure hinge factor (η_f) is determined based on the stiffness calculation equation. Specifically, because the output voltage and the hinge's deflection can be considered proportional, the output energy is approximately in proportion to the squared value of the hinge's deflection:

$$\eta_f = \frac{EWT^3/L^3}{183N/m} \tag{12.17}$$

where L, W, and T represent the length, in-plane width, and out-of-plane thickness of the flexure hinge, respectively. E means the Young's modulus of the material. Nondimensional η_f value is normalized as 1 by dividing the stiffness of the flexure hinge.

Second, the required input energy to be stabilized at a high energy orbit can be calculated as the sum of U_T and U_i. Hence, the energy conversion factor is determined as:

$$\eta_e = \frac{D_s^2}{U_T + U_i} \tag{12.18}$$

where D_s is the displacement between the two potential energy traps (Fig. 12.1).

Table 12.3 Performance comparison versus previous designs of structure-based bistable vibration energy harvester.

Performance	Previous work (Sharpes et al., 2015)	Previous work (Zhang et al., 2018)	This work
Size of one side's flexure hinge (L × W × T)	$36 \times 8 \times 0.6$ mm^3	$50 \times 10 \times 0.1$ mm^3	$70 \times 16.5 \times 1.2$ mm^3
Material of beam	Copper–beryllium	Polyethylene terephthalate	ABSplus-P430
Young's modulus of material of beam (E)	147 GPa	2.95 GPa	2.2 GPa
Flexure hinge factor (η_f)	29.4	1.29×10^{-3}	1
Energy conversion factor (ηe)	4.55×10^{-4} m^2/J	0.16 m^2/J	3.13×10^{-2} m^2/J
Figure of merit ($\eta_{f\eta e}$)	1.34×10^{-2} m^2/J	2.09×10^{-4} m^2/J	3.13×10^{-2} m^2/J
Simulation results			
Magnitude of the potential energy traps (UT)	25 mJ	0.23 mJ	0.46 mJ
Displacement between potential energy traps (D_s)	3.4 mm	18 mm	4.11 mm
Experimental results			
Weight of proof mass (m)	12 g	10 g	65 g
Compression length of preloading	0.08 mm	3 mm	0 mm
Input	20 m/s^2; 38 Hz	16.81 m/s^2; 10 Hz	4.9 m/s^2; 20 Hz
Output	Open-circuit RMS voltage: 42 V	Average power: 313 μW	Average power: 5.87 μW
Frequency range over 30% peak value of average output power	25–40 Hz	–	16–20.4 Hz
Predicted inconsistent energy in gravity	0.40 mJ	1.77 mJ	0.08 mJ

RMS, Root mean square.

Then, the figure of merit is determined as the product of the flexure hinge factor (η_f) and the energy conversion factor (η_e).

12.4.3.2 Performance comparison

The performances of the proposed structure-based bistable vibration energy harvester are compared with those of the previous structure-based bistable vibration energy harvesters in the literature. The results are listed in Table 12.3. Compared with the previous designs of structure-based bistable vibration energy harvester (Andò et al., 2014; Xu et al., 2013), the proposed structure-based bistable vibration energy harvester has less inconsistency when the excitation is imposed in the direction of gravity.

When the excitation is applied in the direction of gravity, the proof mass of the structure-based bistable vibration energy harvester cannot be flexibly selected. Otherwise, bistability will disappear because the predicted inconsistency in the direction of gravity is much larger than the magnitude of the potential energy traps (such as in the previous work) (Andò et al., 2014). Referring to the simulation results in Table 12.3, the magnitude of the potential energy traps of the previous design in Andò et al. (2014) is much larger than others. It means that the design is more difficult to skip between the two energy wells. This is also revealed by the experimental results, which indicate that the required input acceleration to surmount the potential barrier in the previous work (Andò et al., 2014) is much larger than others. By contrast, the proposed energy harvester has a higher figure of merit than those in previous work (Andò et al., 2014; Xu et al., 2013). It represents a more suitable design for energy harvesting dedicated to the direction of gravity.

Because the discrepancy may also be induced by different materials and dimensions of the used piezoelectric films, a more comprehensive comparison needs to be carried out in future by considering more issues that influence the harvester's performance. The proposed energy harvester can be applied to harvest vibration energy in the direction of gravity. For example, it is applicable as an energy harvesting floor to scavenge vibration energy induced by walking people, passing vehicles, and so on. Considering that the direction with an inclination angle to vertical direction also has the effect of gravity, the proposed structure-based bistable vibration energy harvester can also be used to harvest energy in other than a vertical direction. In the future, different piezoelectric polymers or ceramics will also be used to test the performance of the demonstrated system.

12.5 Conclusion

This chapter presents the design and verification of a structure-based bistable vibration energy harvester dedicated to vibration energy harvesting in the direction of gravity. The flexure-based mechanism is designed by integrating the BCM, PVDFs, and proof mass. Based on the established analytical model, a parametric study was carried out to evaluate the influence of each structural parameter on the BCM performance. A simulation study was performed to predict the output performance of the structure-based bistable vibration energy harvester. Experimental results showed that the structure-based bistable vibration energy harvester coincided with the design goal. The maximum output power reached 5.87 μW at 20 Hz under a base excitation of 0.5 g with a resistance of 0.1 MΩ, and the frequency range with over a 30% peak value of average output power was around 16–20.4 Hz. In addition, a figure of merit was proposed for performance comparison with existing designs. Comparison results revealed that the structure-based bistable vibration energy harvester was more suitable for harvesting energy in the direction of gravity. In future work, the performance of the structure-based bistable vibration energy harvester needs further improvement by manufacturing with higher machining accuracy and the use of different piezoelectric materials.

References

Andò, B., Baglio, S., Bulsara, A. R., & Marletta, V. (2014). A bistable buckled beam based approach for vibrational energy harvesting. *Sensors and Actuators A: Physical, 211*, 153–161. https://doi.org/10.1016/j.sna.2013.12.027

Cottone, F., Gammaitoni, L., Vocca, H., Ferrari, M., & Ferrari, V. (2012). Piezoelectric buckled beams for random vibration energy harvesting. *Smart Materials and Structures, 21*(3). https://doi.org/10.1088/0964-1726/21/3/035021

Gu, L. (2011). Low-frequency piezoelectric energy harvesting prototype suitable for the MEMS implementation. *Microelectronics Journal, 42*(2), 277–282. https://doi.org/10.1016/j.mejo.2010.10.007

Harne, R. L., & Wang, K. W. (2013). A review of the recent research on vibration energy harvesting via bistable systems. *Smart Materials and Structures, 22*(2). https://doi.org/10.1088/0964-1726/22/2/023001

Howell, L. L., Magleby, S. P., & Olsen, B. M. (2013). Handbook of compliant mechanisms. In *Handbook of compliant mechanisms*. John Wiley and Sons. https://doi.org/10.1002/9781118516485

L G H, S., E, K., D, P., P, L., & P, E. (2014). Modelling and experimental verification of more efficient power harvesting by coupled piezoelectric cantilevers. *Journal of Physics: Conference Series*, 012098. https://doi.org/10.1088/1742-6596/557/1/012098

Li, P., Liu, Y., Wang, Y., Luo, C., Li, G., Hu, J., Liu, W., & Zhang, W. (2015). Low-frequency and wideband vibration energy harvester with flexible frame and interdigital structure. *AIP Advances, 5*(4). https://doi.org/10.1063/1.4919711

Moro, L., & Benasciutti, D. (2010). Harvested power and sensitivity analysis of vibrating shoe-mounted piezoelectric cantilevers. *Smart Materials and Structures, 19*(11). https://doi.org/10.1088/0964-1726/19/11/115011

Piezo film sensors technical manual.(1999).

Sharpes, N., Abdelkefi, A., & Priya, S. (2015). Two-dimensional concentrated-stress low-frequency piezoelectric vibration energy harvesters. *Applied Physics Letters, 107*(9). https://doi.org/10.1063/1.4929844

Upadrashta, D., & Yang, Y. (2018). Trident-shaped multimodal piezoelectric energy harvester. *Journal of Aerospace Engineering, 31*(5). https://doi.org/10.1061/(ASCE)AS.1943-5525.0000899

Xu, Q. (2018). Micromachines for biological micromanipulation. In *Micromachines for biological micromanipulation* (pp. 1–226). Springer International Publishing. https://doi.org/10.1007/978-3-319-74621-0

Xu, C., Liang, Z., Ren, B., Di, W., Luo, H., Wang, D., Wang, K., & Chen, Z. (2013). Bi-stable energy harvesting based on a simply supported piezoelectric buckled beam. *Journal of Applied Physics, 114*(11). https://doi.org/10.1063/1.4821644

Zhang, X., Gao, S., Li, D., Jin, L., Wu, Q., & Liu, F. (2018). Frequency up-converted piezoelectric energy harvester for ultralow-frequency and ultrawide-frequency-range operation. *Applied Physics Letters, 112*(16). https://doi.org/10.1063/1.5022496

Index

'*Note*: Page numbers followed by "f" indicate figures and "t" indicate tables.'

R

U

V

W

FSC
www.fsc.org
MIX
Papier | Fördert gute Waldnutzung
FSC® C083411

Kolibri 360 GmbH
Ferdinand-Jühlke-Straße 7
99095 Erfurt, Deutschland
produktsicherheit@kolibri360.de